A Cultural History of Sound, Memory and the Senses

The past 20 years have witnessed a turn towards the sensuous, particularly the aural, as a viable space for critical exploration in history and other humanities disciplines. This has been informed by a heightened awareness of the role that the senses play in shaping modern identity and understanding of place, and increasingly, how the senses are central to the memory of past experiences and their representation. The result has been a broadening of our historical imagination, which has previously taken the visual for granted and ignored the other senses. Considering how crucial the auditory aspect of life has been, a shift from seeing to hearing past societies offers a further perspective for examining the complexity of historical events and experiences. Historians in many fields have begun to listen to the past, developing new arguments about the history and the memory of sensory experience. This volume builds on scholarship produced over the last 20 years and explores these dimensions by coupling the history of sound and the senses in distinctive ways: through a study of the sound of violence, the sound of voice mediated by technologies and the expression of memory through the senses. Though sound is the most developed field in the study of the sensorium, many argue that each of the senses should not be studied in isolation from each other, and for this reason, the final section incorporates material which emphasizes the sense as relational.

Joy Damousi is Professor of History at the University of Melbourne.

Paula Hamilton is adjunct Professor of History at University of Technology, Sydney.

Routledge Studies in Cultural History

For a full list of titles in this series, please visit www.routledge.com

A Cultural History of Sound, Memory and the Senses

Edited by Joy Damousi
and Paula Hamilton

Routledge
Taylor & Francis Group

LONDON AND NEW YORK

First published 2017 by Routledge

2 Park Square, Milton Park, Abingdon, Oxfordshire OX14 4RN

52 Vanderbilt Avenue, New York, NY 10017

Routledge is an imprint of the Taylor & Francis Group, an informa business

First issued in paperback 2019

Library of Congress Cataloging in Publication Data
A catalog record for this book has been requested

ISBN: 978-1-138-21177-3 (hbk)
ISBN: 978-0-367-26409-3 (pbk)

Typeset in Sabon
by Apex CoVantage, LLC

Contents

Figures

Introduction
Leaning In

Joy Damousi and Paula Hamilton

The American lyricist Ira Gershwin had a sensitive ear for the sounds of the modern world. In 1932 he listed in his diary: 'Heard in a day: An elevator's purr, telephone's ring, telephone's buzz, a baby's moans, a shout of delight, a screech from a "flat wheel", hoarse honks, a hoarse voice, a tinkle, a match scratch on sandpaper, a deep resounding boom of dynamiting in the impending subway, iron hooks on the gutter.'[1]

The publication of this book makes a contribution to the scholarly conversation about the role of the senses in historical research. The past 20 years has witnessed a turn towards the sensuous, particularly the aural, as a viable space for critical exploration in both history and other humanities disciplines. This has been informed by a heightened awareness of the role that the senses play in shaping modern identity and understanding of place, so evocatively described by Ira Gershwin's daily catalogue of sounds above, and increasingly, how the senses are central to the memory of past experiences and their representation. The result has been a broadening of our historical imagination, which has previously taken the visual for granted and ignored the other senses. Considering how crucial the auditory aspect of life has been, a shift from seeing to hearing past societies offers a further perspective for examining the complexity of historical events and experiences.

The result can be transformative in how we understand and interpret the past. The aim of this volume is 'leaning in' metaphorically to hear more closely the senses of the past through engagement with sources generated at the time, or by remembering at a distance from the time of their experience, which provides us with a means of access to the richness of a world no longer inhabited. Yet this field does so much more than imaginatively re-create the historical landscape. The senses also provide us with tools to explore different aspects of the past hitherto unrecorded or unknown, and to examine afresh a partly known past through a different lens. The anthropologist Constance Classen noted that the investigation of the sensory worlds of past eras should not merely describe the range of sounds and smells that existed at a particular time, as evocative as that might be, but should uncover the meaning that those smells and sounds had for people.[2]

In other words, we should be engaged in not simply a description of many past experiences through the senses, but our work should be capable of interpreting the meaning of those experiences, in particular those that make little sense if only known through the visual. For historian Mark Smith, sensory history is less a 'field' of inquiry and more a habit of thinking about the past, 'a way of becoming more attuned to the past.'[3] When we think about the senses, we are exposed to 'subliminal histories' that may have always been there, but we are now choosing a different path for understanding their meaning.

In general, we have found that the study of the senses asks us to think profoundly again about the nature of historical experience. It is testimony to the maturity of the sensory studies field that a journal, *Senses and Society*, together with various handbooks and readers, have emerged which stake a claim to its significance, chart its trajectories across time and space and push its boundaries.[4] While sound studies in particular has emerged as a major academic field in recent times, much of this material remains ahistorical or focused on technological advances of sound. Historians, anthropologists and cultural theorists have begun to engage with the challenges of examining the history of sound and the senses across time and place, though there remains a lacuna in drawing out these connections to the ways in which they emerge in different ways within a variety of historical contexts.[5] It was within this context and our awareness that one needs an interdisciplinary approach to the topic that we organized a symposium in July 2014 on 'Sound, memory and the senses' at the University of Melbourne. The chapters in this volume are drawn largely from this symposium.

Essays in this book address the senses in two principal ways: some focus on the history of the politics and technologies of sound; others instead ask questions about how the new theoretical work on the senses can inform the research and writing of history. Where ours departs from existing work is first in the methodological use of oral histories by many of the participants. Recently, the anthropologists Douglas Howes and Constance Classen, key figures in charting the sensory studies fields, have argued that anthropologists have an edge over historians because 'through participant observation, they can have firsthand experience of the sensory lives of the people they study.'[6] Historians, they say, have to 'make do' with the usual written sources, visual images and material culture except in more recent periods with 'film and sound recordings.' In fact, historians have been more epistemologically adventurous than this by exploring the sensory possibilities of oral history interviews done either by themselves or others before them, who have been able to explore the senses and historicize them through oral remembering as far back as the beginning of the twentieth century.

Second, there is also little work on memory and the senses, beyond their role as mnemonic triggers, perhaps because it engages the physiological basis of cultural effects. Bruce Johnson, in his essay for this volume, warns against treating the senses as merely a cultural phenomenon, as if they had no basis in the materiality of the body; and there are certainly some who write in this field without acknowledging the crucial role of memory and emotion in the articulation of their existence. Aside from the tacit, all we have of historical experience is known through representation or writing afterwards, recorded sound and image and oral histories. Although always relational, smell, touch and taste are embodied and cannot be separated from the person, whereas like vision, sound responds to external stimuli and has been for a century separated by recording. Writing in *Civilization and Its Discontents* (1930), Freud was very aware of the impact on our memories of the new technologies:

> In the photographic camera we find an instrument which retains the fleeting visual impressions, just as the gramophone disc retains the equally fleeting auditory ones . . . both are thus materialisations of the power of recollection, of memory.[7]

But neither of course could he see how recorded sound became central to the transformation of everyday life—how people found ways of expressing themselves, important dimensions of their personal lives and the memory of their experiences through the recorded auditory and new soundscapes.

Finally, the volume builds on the substantial scholarship that explores the coupling of the history of sound and the senses in distinctive ways: through a study of the sound of violence, the sound of voice mediated by technologies and the expression of memory through the senses. Though all of the senses have received separate book-length treatment by scholars, many argue that the senses should not be studied in isolation from each other, and for this reason our final section incorporates material which emphasizes the senses as relational.

A Cultural History of Sound, Memory and the Senses begins in Chapter 1 with a survey of the interdisciplinary field of sound studies by Bruce Johnson, who identifies the key elements and aspects of this body of work. This is a necessary beginning to the collection, for it navigates readers through the themes which have defined the field. The book is then divided into three sections, each addressing one theme of the relationship of sound to cultural phenomena.

Part I of the volume addresses the ways in which the production of voice has defined cultural experience across the nineteenth and twentieth centuries. In Chapter 2, Henry Reese explores the introduction of that quintessential nineteenth century invention—the phonograph—and positions its emergence within the mechanized sounds of the period. Reese focuses on the rise of a particular vernacular sound, which was both transnational and particular to the colonial landscape. By the 1930s, radio had superseded the phonograph to become the medium of modernity. Both Chapters 3 and 4 focus on the influence of radio in the 1930s and the 1940s, to explore the power of this medium that was defined and shaped by speech and the voice. Jennifer Bowen explores this theme in the context of the rise and influence of Australian radio and the enormously popular book readings at this time. Her chapter poses questions about listening and whose voice is being heard and how. Following on from this theme, David Goodman focuses on the voice of the aviator Charles Lindbergh, and his charismatic speeches against America's entry into World War II on the radio and at mass public events. The response to Lindbergh's speeches and the emotions in his voice direct us to consider the connection between voice, emotion and memory. Chapter 5, Kate Darian-Smith's exploration of voices and noise in the classroom during the 1960s and 1970s, highlights shifting attitudes to acceptable levels of sound in pedagogical practice, especially in relation to children's learning.

Violent acts are invariably accompanied by distinctive and memorable sounds. Those who are victims, witnesses or perpetrators of violence invariably remember the acts through sound. Part II explores the connection between sound and violence by highlighting how sound often contributes to defining the very act of violence. In Chapter 6, Penny Russell examines the sounds around a violent rape in 1832, such as insults, intonations and screams. The testimonies in the subsequent court case reveal the way in which the incident was recalled by witnesses through these sounds. In the same period—early colonial Australia—the emergence of gunfire reflected a new form of aggressive individualism. In Chapter 7, Diane Collins explores the link between settler colonialism and violence, in the context of acceptable and unacceptable forms of sound. In Chapter 8, Joy Damousi explores the impact of sound and violence in two contrasting contexts. The memory of the sound of the Dresden bombings remains with survivors, while the terror embedded in everyday life through silence and sound during the occupation of Paris was a matter of life or death. In Chapter 9, Vannessa Hearman focuses on the 1965–66 methods of repression and violence in Indonesia and how the sounds these generated are indelibly imprinted on the minds of survivors. Their memories are often of the lingering sounds of killing. Less than a decade later, the use of torture following Pinochet's coup in Chile shocked the world. Peter Read discusses this in Chapter 10, analysing the distinctive, haunting and eerie sounds of mass executions. As well as sound, the silences too shape the powerful act of remembering.

In Part III, the focus moves to exploring how the memory of sound has an integral relationship with the other senses, especially smell and touch. In Chapter 11, Lisa Murray explores the broader sensory experiences of the working class, industrial suburbs of inner city Sydney, and the memories of living in this crowded environment, as well as the politics of environmental change. Chapter 12 develops a further aspect to memory and the senses. Paula Hamilton considers the sensory environment, especially smell and touch, of working as a domestic servant in what is traditionally known as the private space of the home and illuminates the changing nature of intimate domestic relations in the first half of the twentieth century. Moving to consider the antithesis of the urban—the botanical environment—in Chapter 13 John Ryan examines the memory of smells through botanical memories of conservationists, environmentalists and wildflower enthusiasts. In a more contemporary setting, in Chapter 14 Lauren Istvandity draws on the power of music to develop a sound autobiography connecting personal memories of place and sound. Finally, in Chapter 15, Emma Dortins concludes the volume by exploring the relationship between two leading Australian women of letters: the poet Judith Wright and Indigenous writer and poet Kath Walker (Oodgeroo Noonuccal), through the power of visual metaphor and racialized sound.

Collectively, these essays explore how the senses illuminate the historical and what sources are available to read the senses of the past through the material and memory. The latter is less tangible, but no less powerful. But these essays also point to the value of an interdisciplinary approach to position the senses at the centre of historical inquiry. They point to a variety of approaches and methodologies in adopting this focus and illuminate a multiplicity of ways to write and research histories of the senses. Such a richness of approaches and the application of a variety of methodologies points to innovative ways forward for future research in sound and sensory studies.

Notes

1 Rosenberg 1991, p. 31.
2 Classen 1993, p. 6.
3 Smith 2007, p. 1.
4 See Erlmann 2004; Sterne 2012; Pinch and Bijsterveld 2013.
5 For recent studies, see Morat 2014; Smith 2014; Street 2015.
6 Howes and Classen 2014, p. 4.
7 Freud 1930, pp. 37–39.

References

Classen, C., 1993. *Worlds of sense: Exploring the senses in history and across cultures*, New York: Routledge.
Erlmann, V., ed., 2004. *Hearing cultures: Essays on sound, listening and modernity*, London: Bloomsbury.
Freud, S., 1930. *Civilization and its discontents*, J. Strachey trans., *The standard edition of the complete psychological works of Sigmund Freud, Volume XXI (1927–1931)*, and published as a single volume, New York: Norton, 1961.
Howes, D., and Classen, C., 2014. *Ways of sensing: Understanding the senses in society*, London: Routledge.
Morat, D., ed., 2014. *Sounds of modern history: Auditory cultures in 19th and 20th century Europe*, New York and Oxford: Berghahn Books.

Pinch, T., and Bijsterveld, K., eds., 2013. *The Oxford handbook of sound studies*, Oxford: Oxford University Press.

Rosenberg, D., 1991. *Fascinating rhythm: The collaboration of George and Ira Gershwin*, New York: Penguin Books USA.

Smith, M., 2007. *Sensing the past: Seeing, smelling, tasting and touching in history*, Berkeley, CA: University of California Press.

Smith, M., 2014. *The smell of battle, the taste of siege: A sensory history of the Civil War*, Oxford: Oxford University Press.

Sterne, J., ed., 2012. *The sound studies reader*, London: Routledge.

Street, S., 2015. *The memory of sound: Preserving the sonic past*, Oxford and New York: Routledge.

1 Sound Studies Today

Where Are We Going?[1]

Bruce Johnson

The Contemporary Soundscape

In December 2009 a young student attending his first university party in London found himself crowded against a bass speaker. He said to a friend, 'My heart feels funny. I think the bass is affecting me. Oh God, I feel very weird. My heart is beating so fast.' Minutes later he collapsed and died. Cause of death was recorded as Sudden Arrhythmic Death Syndrome (SADS), and, according to a medical spokesperson, possibly attributable to 'a lot of loud noise.'[2] This is a dramatic recognition of the unprecedented power of the contemporary everyday soundscape. In specific terms, I believe the medico was wrong about volume as the fatal parameter, but that it was pitch,[3] but for the moment we can say that it certainly underscores the contemporary importance of sound studies.

The power of sound is recognized as far back as human records take us. The Greek Orpheus and the Finnish Väinämöinen are two of many characters from mythology who, at a central moment in their narratives, become engaged in lethal singing contests. In Biblical times, Jericho was brought down by sound, and the Biblical God is always a sonic presence and almost never seen. In Elizabethan England, travellers arriving from the provinces to London heard the city before they saw it, and Londoners themselves complained of repetitive music in public spaces, a cry that echoes through to the present.[4]

But with the emergence of modernity, the soundscape has changed so radically that we can chart the history of modernity specifically through the changing politics of the soundscape, the changing status of noise, sound and ways of conceptualizing sonicity. One unequivocally distinguishing feature of modernity is the changing soundscape. The modern soundscape is distinguished by the level and complexity of constructed sound. The sounds of nature are still with us in pretty much the forms they have always been. But the sounds constructed by human beings have transformed the acoustic environment. From the late nineteenth century, two areas of technological development produced an unprecedented change in that environment: the sounds of technologies and technologized sounds.[5] For the first of these we can blame industrialization, which produced machinery that drowned out the voices of its human operators, and which was documented extensively by Victorian novelists from Charles Dickens through Elizabeth Gaskell to Thomas Hardy. We can conveniently date the advent of the second, technologized sound, from 1877 with the patent issued for sound recordings, from which emerged the prolific and disparate world of sounds mediated by modern technologies: radio, movies, television, telephonics, electronically amplified musical instruments—to the world of sound digitization ranging from new recording technologies to personal stereos.

World War I provides a useful watershed moment for both of these developments. Technologized sound became pervasive in many forms, from military communications systems to the portable phonograph, which a surprising number of servicemen were able to haul about with them. And it was also the war in which the single most frequent category of trauma reported by men in the trenches was the noise of military technology, a flood in which all sense of individual identity was drowned.[6] World War I utterly transformed the sonic imaginary, ushering in a century in which everyday life was inundated by constructed sound, from the almost subliminal hum of mains electricity and air conditioners, road and rail traffic, to the impulse sounds of machinery and piped music, and the shriek of jet aircraft. As Futurist Luigi Russolo, in *The Art of Noises*, 1913, declared extravagantly, in the nineteenth century, 'with the invention of the machine, Noise was born,'[7] and while he and his circle embraced it as the sound of the future, it would become one of the greatest threats to human welfare of its time. At the end of the twentieth century, a Commission of the European Communities reported: 'Present economic estimates of the annual damage in the EU due to environmental noise range from EUR 13 billion to 38 billion. Elements that contribute are a reduction of housing prices, medical costs, reduced possibilities of land use and cost of lost labour days.'[8]

Sound Studies: A Historical Overview

The advent of sound studies is thus very much a product of the material culture of the modern world. The early history of soundscape studies reflects the growing awareness of what is distinctive about the modern soundscape. R. Murray Schafer's 1977 benchmark study, *The Soundscape: Our Sonic Environment and the Tuning of the World*, includes essays on the history of the soundscape with comparisons between the pre-modern and the modern sonic environments. His foundational work in the 1960s was conducted with the World Soundscape Project, which included Hildegarde Westerkamp and Barry Truax, whose *Handbook of Acoustic Ecology* of 1978, was, I believe, the first of its kind. One of the earliest collections outside the World Soundscape Project was the 1994 *Soundscapes: Essays on Vroom and Moo*, edited by Finnish scholar Helmi Järviluoma who, with Greg Wagstaff, also published the 2002 book *Soundscape Studies and Their*

Methods. Järviluoma has built on the foundations set down by the World Soundscape Project and Schafer, including by revisiting and extending his original field locations to examine changes in the decades since the original World Soundscape Project. The results were published in 2009, as *Acoustic Environments in Change*, edited by Helmi Järviluoma, Meri Kyto, Barry Truax, Heikki Uimonen and Noora Vikman, and also reprinting Schafer's 1977 *Five Village Soundscapes*. Other significant scholars range across disciplines, as for example cultural historian Alain Corbin, ethnomusicologist Steven Feld, science historian Jonathan Sterne and literary historian John Picker.[9] Significant sound study institutions include the founding of the World Forum for Acoustic Ecology in 1993, which now has its own journal and annual conferences; and the establishment of CRESSON sound research centre in Grenoble, France, in 1998; there are similar centres including those in Trondheim, Norway and Lund, Sweden.

A number of the names I have mentioned remind us that this tradition seems to be stronger in some cultures than others. The World Soundscape team was Canadian; while Järviluoma, Kyto, Uimonen and Vikman are all Finnish researchers. There is also a very active sound studies community in Japan. Different communities have very different levels of sensitivity to sound, and this usefully foreshadows the central argument I want to make: that is, the importance of material cultures in our approaches to this field. As we bring sound studies into convergence with materiality, it is poised to become the most productive driver of a fundamental paradigm shift that is emerging in the study of social practices.

Acoustemology

Steven Feld coined the term 'acoustemology,' by which he means 'knowing through sound.'[10] It reminds us that the field of sound studies raises episte mological issues.

In the anglophone tradition, authority is embodied in information and knowledge conceived in terms of a visual order: perspective, vision/visionary, envisage/envision, point of view, discover, disclose, observation, speculation, illustration, demonstration, reflections, insights, second sight, revelation, theory (from the Greek word for 'spectacle'). Our understanding of the modern era is based on the 'Enlightenment,' we study to 'cast light on,' to 'light the path.' Even in these first few pages of a chapter that is trying to favour sonic modes, I have used the following visual metaphors for knowledge and information: recognition, foreshadow, chart, distinguish, documented, reflects, examine. English is a language in which a musical performance is likely to be called 'a reading,' and the most respected music is scored for the eye. It is a language in which we often announce an idea with the words 'apparently' or 'it appears that,' and we ask our interlocutor 'Do you see what I mean?,' where a doctor about to probe and listen to the body begins by saying 'Let's take a look at you.'

By contrast, sonically embodied knowledge has become for us the object of suspicion and derision since the ascendancy of the scientific conscious-ness. In English, the complement of 'Seeing is believing' is the warning 'Don't believe everything you hear.' It is noteworthy how many of the fol-lowing terms for aurally transmitted information arose, or forfeited their cultural capital, during the seventeenth and eighteenth centuries: hearsay, gossip, tittle-tattle, sounding off, chatter, whining or moaning, Chinese whispers, rumour, lip service, scolding, nagging, blab, babble, prate, prattle. The English language does not have much positively charged space for audi-tory information: in his benchmark English *Dictionary*, Samuel Johnson explicitly declared that he would include no words unless they were in print, thus in effect banishing illiterate oral cultures from the anglophone com-munity.[11] To be a meaningful member of the community of the English was to be able to read.

There is nothing natural or universal about this scopocentricity. I dwell on it for a moment because the point needs to be made that sound stud-ies require a particular kind of 'knowing' that does not come easily to all cultures. The cultural study of sound is not simply a shift of interest from the visual to the sonic. It requires the cultivation of a particular phenom-enology and its own conceptual modelling. It requires ideally nothing less than a change in epistemology. I became most strongly aware of this when I started to learn a language from a completely different language group, Finno-Ugric as opposed to Indo-European. In my case it was Finnish, and I found this to be a more hospitable medium for talking about sound because the language and culture are more sonocentric. This is related to history (oral tradition), climate (snowscapes), topography (deep forests) and demography; for example, half of the population of Finland lives no more than 200 metres from a forest and the rest not more than about 3.5 kilometres away.[12] Also implicated is something as elusive as national identity, which for the Finns is associated with silence, and polite citizen-ship with taciturnity. Like Japanese, the sonicity of the culture is inscribed in its language. Like all languages close to their oral sources, all Finnish words are sounded exactly as they are spelled; that is, there is a closer link between sound and orthography than in English. A few of many illustra-tions of its sonocentricity could include the way time is imagined: the timepiece that we call a 'watch' in English is called '*kello*' in Finnish, which is also the word for 'bell.' As in so many languages, proverbial wisdoms are significant, and in Finnish these are notably sonic in orienta-tion, including: 'Respect the deep voice of experience' and '*Sitä kuusta kuuleminen jonka juurella asunto*'—meaning that wisdom lies in listening to the roots of the fir tree next to your house. In English, we are likely to greet someone with the words 'Nice to see you.' In Finnish, a common formal greeting is '*Mita kuluu*': 'What do you hear?' And in fact the word '*kuluu*' also represents the notion of a community—a group that hears the same sounds. To be a meaningful member of the community of the Finns is to be able to hear: '*Minä kuulun tänne*' ('I hear in this place') means 'I belong here.'

There is a significant relationship between sound studies and the mediating influences of language itself. In brief, the English language happens not to be the language best equipped to discuss the complexities of sonic phenomenologies. I have suggested that this is partly because of its visual orientation, but also because modern English prose was largely shaped by the scopocentric scientific programme of the Royal Society. It 'thinks' best in terms of sharply visualized conceptual distinctions, which sonic phenomena often challenge, as I shall argue in a little more detail later.

Sound Studies and Cultural Theory

Over the last decade or two, one of the most conspicuous developments in sound studies has been its increasing appropriation by more general cultural theory. If we want to understand where sound studies is today, its relationship with cultural theory requires some discussion.

Cultural theory in anglophone academia emerged from the growing recognition of the ideologies underpinning scholarly 'objectivity.' My first disciplinary base was literary studies, where the ascendancy of theory was a reaction against what was basically the 'practical criticism' of I. A. Richards and his generation. Their objectification of the text as the exclusive analytical focus came to be regarded with justification as ahistorical, based on such articles of faith as 'universal values' in human behaviour and objective literary standards as embodied in the canon. They denied, in effect, the importance of cultural and historical context in the experience of literature.

Cultural theory developed as a reaction against this, and from the 1970s, it became intellectually omnivorous, engulfing established disciplines including literary studies, history, anthropology and musicology.[13] Cultural theory became an imperialist in the intellectual landscape, colonizing indiscriminately. Names like Bourdieu and Foucault, and a little later their post-structuralist successors like Derrida and Deleuze, were just as likely to underpin discussions of the cultural history of music as of medicine, on the assumption that one size fits all.

In the early twenty-first century, cultural theory has taken up sound studies as yet a further demonstration of its amphibious versatility. I see this as the current impasse within which the most faddish manifestations of sound studies are now trapped. The reason that I use that image is that there are two aspects of cultural theory as it flourished from the late twentieth century that are in irreconcilable tension with the basic phenomenology of sound.

First: Especially in its anglophone manifestations, cultural theory is fundamentally framed by a scopic epistemology. The dominant discourses of cultural theory, even when deployed in the study of sound, remain colonies in the empire of the visual. This is especially evident in the theoretical frames within which cultural theory has worked, which were not derived from sonic but from visual epistemologies. I illustrate by reference to one of the most influential cultural theorists of the late twentieth century, Fredric Jameson. One of his benchmark books was *The Political Unconscious*, from 1981, which began with the announcement: 'This book will argue the priority of the political interpretation of literary texts. It conceives of the political perspective as . . . the absolute horizon of all reading and all interpretation.'[14] This way of conceptualizing knowledge is firmly in a tradition that can be traced back to Francis Bacon, who introduced *The Great Instauration* by declaring that 'I . . . dwelling purely and constantly among the facts of nature, withdraw my intellect from them no further than may suffice to let the images and rays of natural objects meet in a point, as they do in the sense of vision.'[15]

An over-arching tradition linking the early modern and the postmodern is disclosed in the unquestioned reliance on visual metaphors as a way of conceptualizing knowledge: images, rays, vision, perspectives, horizons. This 'visualization' of knowledge and its link with power have remained deeply embedded in a range of cultural discourses, from Michel Foucault on mechanisms of punishment and control to Laura Mulvey on film.

Gradually, as I have moved more into the study of sonic culture, I have become aware of the constraints imposed on our understanding of culture by these visual tropes. Those constraints are increased by the fact that the scopic frame has been largely invisible, subconscious and therefore uninspected. In addition, however, Jameson is plain wrong, because in our negotiations with the world out there, there is a more encompassing mediation than the 'political': it is the body. Before we can begin to conduct any political interpretation of the world we have to physically experience it—read it, hear it, touch it, taste it, smell it. Without the body, there can be no community of knowledge, or indeed of any kind. So obvious—yet so long overlooked.

Jameson thus also exemplifies the other major problem with mainstream cultural theory: it overstates the culturalist basis of social practices. In doing so, for decades it attached an excessive importance to ideation and cognition as the drivers of behaviour and belief. Cultural theory, when positing the *cultural* construction of so many aspects of identity and its relationships with society, has generally written the body out of its discourses. Phenomena such as affect, ideology, semiotics—the whole realm of meaning which is central to the study of culture and cultures—have been lifted off the bone, so to speak, and identified in ideational terms. Elizabeth Wilson commented that feminist studies think of 'bodily transformation ideationally and symbolically, without reference to biological constraints. That is, to think about the body as if anatomy did not exist.'[16]

In its fixation on culture as the key to everything, cultural theory, at least until very recently, severed itself from biology and more generally from the banal, vulgar realities of materiality; and wherever it intersected with the study of sound, it tended to detach sound studies from physiology. That is, it showed little recognition of what is distinctive about sonic experience, of the distinctive relationship between sound, memory and the senses. Sound simply became another, increasingly fashionable, 'hook' for cultural analysis. I have over the last few years attended conferences on soundscapes hosted by departments of literary studies. For the most part, they consisted of papers that pointed out how often a particular literary work referred to sound. They were frequently not studies of sound at all, but studies of literary imagery, from which the distinctive phenomenologies of hearing were completely absent. I want to show shortly just how crippling this 'disembodiment' has been to culturalist approaches to the understanding of sound.

Sound Studies Now

Transdisciplinarity

In the twenty-first century, the status of sound studies has increasingly been validated by a proliferation of compilations and compendia, which themselves point to an extensive body of literature. In addition to other studies mentioned throughout this discussion, examples include Michael Bull and Les Back, eds, *The Auditory Culture Reader* (Oxford and New York: Berg, 2003); Jim Drobnick, ed., *Aural Cultures* (Banff, Canada: YYZ Books, 2004); Veit Erlmann, ed., *Hearing Cultures: Essays on Sound, Listening and Modernity* (Oxford and New York: Berg, 2004); and Erlmann's *Reason and Resonance: A History of Modern Aurality* (New York: Zone Books, 2010); Hillel Schwartz's *Making Noise: From Babel to the Big Bang & Beyond* (New York: Zone Books, 2011); Jonathan Sterne, ed., *The Sound Studies Reader* (London and New York: Routledge, 2012) and Michael Bull, ed., *Sound Studies*, also from Routledge; as well as more specialized texts such as Emily Thompson's *The Soundscape of Modernity: Architectural Acoustics and the Culture of Listening in America, 1900–1933*, (Cambridge Mass. and London, MIT Press 2004); Mark M. Smith, ed., *Hearing History: A Reader*, (Athens, GA: University of Georgia Press, 2010); Ross Brown, *Sound: A Reader in Theatrical Practice* (Basingtoke: Palgrave Macmillan, 2010) and Tim Crook, *The Sound Handbook* (New York and Abingdon Oxon: Routledge, 2012).

One pattern gradually emerging in this literature is an appreciation of the convergence of the intellectual and the material in auditory cognition. This is exemplified in one of the most recent collections, in which Oxford University Press confers its authority on the field, *The Oxford Handbook of Sound Studies* (Trevor Finch and Karen Bijsterveld, eds, 2012), which recognizes the materiality of hearing through contributions that range through cognitive studies, acoustic engineering, physiology, ecology, geography and neuroscience. As with other such titles, the collection suggests that sound studies now has a multiple tradition. But it is compartmentalized by territorial quarantines, which are sometimes even explicit. The International Community for Auditory Display (ICAD) was established to draw together scientific studies of what it calls 'sonification.' Its first conference, in 1998, was largely devoted to defining its disciplinary boundaries, in the course of which it declared imperially that 'the humanities' in general and music in particular should not participate in what it asserts is an exclusively scientific enterprise.[17]

Yet some of the work coming from ICAD makes it very obvious that they—and all sound studies stakeholders—should talk more to each other. Science and technology studies would benefit from reading for example popular music studies, but also vice versa. Music researchers have often disdained scientific disciplines. I think there is an anxiety that thinking of music as a subject of scientific analysis (as 'sound' or even 'noise') forfeits the aesthetic gravitas that justifies their attentions. They have preferred as their framework a version of cultural theory which is often oddly detached from the material world. Incursions into sound studies from both the humanities and the sciences thus display disciplinary based lacunae. Perhaps one path forward is outside the academy, in industrial and commodity research projects like Objective Evaluation of Interior Car Sound (OBELICS). This was a three-year project launched by the European Union in 1997 with a view to understanding, for commercial reasons, subjective responses to sounds heard inside a car. Such projects are not constrained by disciplinary-based territorialisms, but really are interested in a practical engagement with social practices rather than establishing a lofty distance from them. Automobile manufacturers want their business models to be intimately related to the 'real world' of social practice and consumer desire, unlike the cultural theorists whose agenda is, alas, likely to be intellectual foppery, a 'display' or 'performance.' The engineers in the corporate sector, on the other hand, relentlessly sought to replicate the 'real' in developing their test models.[18]

Unfortunately, while all these approaches have essential things to say about sonic phenomenology, they all too rarely talk to each other. Apart from occluding cross-disciplinary continuities,[19] from the point of view of the humanities, it means that our approach to sound studies has been largely disembodied and dematerialized, a curious situation given the emergence of histories of the senses in which sound studies have been central.

Deaf Spots

The result is 'deaf spots.' That is, the sonic equivalent of blind spots. The sovereignty of the culturalist model as the primary explanation of sonic affect continues to sustain deeply entrenched misapprehensions about sonic experience, and this compromises much of the literature on sonic cultures. There is not a week that goes by without publication of the most strident assertions about sound that are at best debatable and at worst simply wrong—the current controversies regarding wind farms are a topical example. Here is another, from a different discourse, as recent as March 2014. The prestigious and influential *London Review of Books* carried an article on 'The Public Voice of Women' by cultural commentator and Cambridge University Professor Mary Beard. She spoke in particular of the association between a low or deep sound and the projection of power. That association is well understood, especially by cinema composers and sound engineers (think of the *Jaws* theme and the voice of Darth Vader). There is one assertion made in the article, however, that has become deeply entrenched but, expressed so categorically, is simply untrue: 'There is no neurological reason for us to hear low-pitched voices as more authoritative than high-pitched voices.' That is, she is arguing that the association between low registers and power is cultural. It is not, and an approach to the study of sound that refers to the body will make this very clear.[20] And I emphasize: if we are asking 'Where is sound studies today?,' this topical example helps to illustrate what I have called its impasse.

A second current obstacle to the development of sound studies is an uncritical romanticization of the sonic experience. This is most pervasive in studies of popular music and its media. Examples include well-established and otherwise instructive and admirable studies such as Tia De Nora's *Music in Everyday Life* (Cambridge: Cambridge University Press, 2000) and Michael Bull's *Sounding out the City: Personal Stereos and the Management of Everyday Life* (Oxford and New York: Berg, 2000). Perhaps because both of these are groundbreaking studies, they present music as unequivocally empowering. Neither gives attention to the rebarbative physical impacts of exposure to sound, as in the cases of music for aquaerobics in the highly resonant space of an indoor swimming pool, or in Bull's case, the use of personal stereos. Both will produce irreversible hearing damage, and the latter insulates the listener from a soundscape that includes important warning signals about traffic and potential assault. Again, we have a culturalist-based framework that ignores the corporeality of sonic experience. The focus on the socio-cultural power of music has left us ignorant as to the role of the body in the determination of sonic affect, as well as the deeply harmful potential of the contemporary soundscape.

This very sketchy survey of the current state of sound studies opens several lines of inquiry. Two are: What might we learn when we talk across disciplines? What mistakes might we cease making in our understanding of sonic phenomenology and affect?

Sonic Affect and Physiology

Let me illustrate by reference to the materiality and physiology of sounding and hearing. The question of sonic affect is central to the cultural history of sound. But how, in material terms, does hearing itself mediate the meanings of what is 'out there?' As a matter of general knowledge, we know that the human ear, even in pristine condition, cannot pick up all the sounds that are available to be heard, but which other species can hear. How else might the body act as gatekeeper in the interpretation of sound?

Sound produces emotional responses—ranging from happiness through sadness to fear. But that first emotional response is not 'cognitive'—that is, the first response is not processed intellectually, but physiologically, insofar as the distinction is tenable (see further below). It is our body rather than our mind that determines our first interpretative response to a sound. We have no control over it. This is because of how sound is processed. Let us consider the journey from sound source to sonic cognition in two stages: first, from the sound source to the ear; and second, from the ear to the brain.

From the Sound Source to the Ear

The physical space in which we hear sound already begins to lay down the foundations of interpretation: the potential of reverberation, for example, to induce a state of awe, has been recognized since antiquity. Like reverberation, sound whose source we cannot easily locate, such as sound coming from above us or low-frequency sounds, will also set up unbalanced power relations that produce emotions ranging from irritation through anxiety to terror. The CRESSON sound research centre calls this the 'ubiquity effect.'[21] The power of sounds that, for various reasons, are difficult to source and localize, has been understood and deployed especially by dramatists from antiquity through to Shakespeare, and down to film music composers and sound engineers.[22] The reasons that we cannot locate these sources and the anxieties they therefore cause are panhistorical, transcultural and even transpecial: experienced by birds, beasts, fishes and evidently even insects. This is because they are to do with the general physics of sounding and the physiology of hearing, the mediating body. Unlocalizable sounds cause varying levels of anxiety. But what makes them unlocalizable is not culture. It is 'nature'—the way our ears sit on our heads, the way sound strikes those ears.

From the Ear to the Brain

The next stage in the journey of the sound to the interpreting consciousness is from the ear to the brain. In his influential 1980 paper 'Feeling and Thinking: Preferences Need no Inferences,' R. B. Zajonc postulated that 'the very first stage of the organism's reaction to stimuli and the very first elements in retrieval are affective.'[23] Our first response to a stimulus is emotional, and is the outcome of physiological rather than cognitive processes. Neuroscience tells us that this is generated in the amygdala, which controls the release of adrenal steroids, which assist the body in preparing to deal with threatening situations.

The sonic signal arrives at the amygdala and it stimulates those emotional responses. But the sound takes two paths to the amygdala. One is via the auditory cortex, a high level information processor that can sort through our memories to identify the precise nature of the stimulus—'Aha, some music . . . no it's not The Beach Boys' *Pet Sounds*, it's the Beatles' *Revolver*.' But the other path is direct from the ear to the amygdala, called a quick and dirty path. Quick, because that signal arrives first. Dirty, because it has a lower level of discrimination, telling the amygdala just that this is scary, happy, safe, threatening—it is the primary response we make to an auditory stimulus, and you will recognize it in sensations like goose pimples, hair prickling, and like your pets as their hackles rise, it can occur before we are even conscious of the precise cause. It is a primary pre-cultural survival mechanism, preparing us for fight-or-flight. So: there is a primary reaction to a threatening stimulus. It is physiological, involuntary and almost irrevocable. It creates a foundation on which a secondary and cognitively mediated response draws on cultural memory to articulate the precise nature of that threat.[24]

For example: a low noise is heard, producing involuntarily the physical symptoms of alarm. A cognitive process involving cultural memory then puts a face to the threat: marauding horsemen, tanks, a tsunami, an approaching mob. The immediate point is that the primary interpretive response to the sonic stimulus is physiological.

My immediate point is this: clearly, before sound studies start addressing the effects of cultural mediations like ethnicity, gender or interpretative models based on eminent theorists like Lacan or Deleuze, something else transcultural has already taken place at the level of spatiality and physiology to establish an interpretive foundation on which the cultural scaffolding can be constructed. These involuntary affective interpretations I have been describing are physiological, not cognitive, according to the models we deploy. A study of sound production, transmission and reception thus discloses that, first, it is not the judiciously discriminating 'mind' that initially shapes sonic affect. It is what, in that model, we would call the body.

But this raises interestingly awkward questions. If it is the body that lays down the platform for music affect, that means the most fundamental of the mediations is also the most fundamental interpretive mechanism. This raises the radical question: in sonic experience, where do the physical mediations stop and the interpreting 'me' begin? Where is the 'me' in the mediations?

Sound Studies and the Mind/Body Dichotomy

Judith Becker has written that the study of 'Music and emotion . . . dissolves intractable dichotomies concerning nature versus culture, and scientific universalism versus cultural particularism.'[25] When sound studies attend to the body, they are able to challenge categories which are deployed by the hegemonic epistemology. Such categories can be an impediment to understanding hearing, as the work of seventeenth and eighteenth century theorists demonstrates. The theorization of music enjoyed a particular efflorescence from the late seventeenth century in conjunction with the newly formed Royal Society, which designed experiments in music and hearing, including comparisons between tuning by theories based on mathematics (ratios) and tuning by the ear of a professional performer and teacher—an encounter between intellectual theory and physical practice. Rather dismayingly, the living ear failed to detect supposed 'errors' in mathematical tuning of up to one quarter of a semitone.[26] A central puzzle was the ear's ability 'to recognize exact ratios when they are expressed *in sound*, and to tolerate considerable deviation from those ratios.'[27] The 'puzzle' of sound, the disparity between hearing and theory, is a further reminder of its distinctive and rather messy phenomenology and its ability to confound tidier scopocentrically articulated models of analysis.

Sound studies destabilizes a whole array of conceptual models that underpin cultural theory in its dominant forms. These models involve binaries like 'subjective/objective,' 'self/other,' 'culture/nature' and of course 'mind/body.' If these are disturbed, it affects the *structures* of our thinking, which in turn affects what we can think. I want to conclude by pointing towards some of the rich and radical implications of sound studies, if we can allow them to challenge the constraining binaries that are deeply and invisibly embedded in cultural theory and cultural studies.

Studies of sonic affect provide empirical evidence of the fragility of those binaries. The body is as much a filtering mediator of meanings as is culture. Indeed, the very concept of 'mediation' can no longer be taken as a given. It is clear that sonic studies challenge the mind/body model: where does the mediating auditory interface with the human experience begin and end? In sonic processes, where does the material become the cognitive? This question has become increasingly urgent in a number of fields of inquiry, perhaps most explicitly in what is referred to as Extended Mind Theory and Distributed Cognition. These are approaches to the study of cognition that model human thought as 'inextricable tangles of feed-back . . . loops, that promiscuously criss cross the boundaries of brain, body and the world.'[28] Such approaches 'require that traditional boundaries among individual, object, environment, and the social world be redrawn.'[29] In the words of J. Hollan, E. Hutchins and D. Kirsch, while traditional models of cognition 'look for cognitive events in the manipulations of symbols inside individual actors, distributed cognition . . . does not expect all such events to be encompassed by the skin or skull of an individual.'[30] Cognition, that is, is a process most usefully understood as not confined to the brain pan, but also conducted by the body itself.

Clearly, this in turn has radical implications regarding the status of gesture in relation to cognition. Our gestures and the material accessories they engage with are not 'mere' representations of ideas fully thought out in the mind. They are cognition in process. Gesture is cognition. Peer refereed experimental work on gesture has established a number of principles salient to this discussion:

- gesturing improves learning
- while gesture is universal, the specific gestural repertoire is language-bound
- gesture is a mode of cognition, not a representation of it, and in some cases it is the *only* mode of cognition.[31]

Conventionally we think of gestures as ways to represent cognition, as reflections of thoughts that have already been generated in the mind. Extended Mind Theory seeks to test the hypothesis that cognition is not *represented* by these signs and gestures, but is *conducted through them*. That is, the distinction between the cognizing me and the corporeal mediations is a false one. Proscribe the gesture and you extinguish the thinking. People active in a number of otherwise disparate fields, from neuroscience to cultural memory, have already crossed the line into this kind of inquiry, if not in so many words. As in the case of the puzzle about where mind and mediations meet, these approaches have the potential to reconstruct the hierarchical mind/body model which has in so many ways demeaned the role of the body and its interactions with the world, as being of a lower status and even an impediment to the operations of the mind. Thus, for example, Sorbonne musicologist André Pirro, who disdainfully declared to his pupil Jacques Chailley that he did not attend concerts: 'Why listen to music? To read it is enough.'[32]

Approaches to cognition similar to Extended Mind Theory are distributed cognition and cognitive ecology. They give us a far richer understanding of how essential physical gestures and their contexts are to the formation and maintenance of memory and belief. The connection has recently begun to be exploited in the treatment of trauma through an approach known as Eye Movement Desensitization and Reprocessing (EMDR).[33] Gestures are not simply the incidental outward show of an interior cognitive state that can be dispensed with without changing that interior state; they are essential to its formation, they are the things through which thought actually takes place, not just manifestations of thought. A recent study of singing among Icelandic men has argued that voice and gesture do not simply reflect nuanced individual identity, but are actually means by which it is formed.[34] This specifically musical datum is an appropriate way to draw my discussion to its conclusion. It exemplifies the role of physical and auditory gesture in the history of belief, in the formation of memory, of communities and their beliefs.

Emerging theories of extended mind and gestural cognition are re-integrating linkages between inner and outer, between thought, the body and the material world that had been severed by Cartesian dualism. And the most promising path into this field is that of sound studies.

Notes

1 The following essay is in large part an overview drawing on work conducted over about two decades, much of which has been published in books, articles, conference and symposia presentations, the most substantial of which are listed in the course of the chapter.
2 Cited in Johnson 2015, p. 82.
3 In particular, the low frequency sounds coming from the bass speaker. See Johnson 2015, pp. 83–86.
4 See Johnson and Cloonan 2008, p. 37. On the soundscape of Elizabethan London, see. Smith 1999, pp. 52–71.
5 See further Johnson and Cloonan 2008, pp. 49–63.
6 Johnson and Cloonan 2008, pp. 50–56. The point also emerges in fictionalized 'memoirs' of ex-service personnel; see for example Frederick Manning's *The Middle Parts of Fortune*, 2012, especially Chapter XVI, pp. 245–57.
7 Russolo 1986, p. 23.
8 MacNevin 2005.
9 For example: Corbin 1998; Picker 2003; Sterne 2003; Feld 2012.
10 On the background, see http://www.acousticecology.org/writings/echomuseecology.html, accessed 28 July 2015.
11 See further Johnson and Cloonan 2008, pp. 43–44.
12 See http://www.metla.fi/metinfo/monikaytto/lvvi/tietoa-ulkoilusta.htm, accessed 29 July 2015.
13 A parallel argument is set out in the course of Johnson 2013.
14 Jameson 1981, p. 17.
15 Bacon 1960, p. 13.
16 Wilson 2004, p. 69.
17 Pinch and Blijsterveld 2011, pp. 253, 254–55, 258, 260.
18 See for example Pinch and Bijsterveld 2011, pp. 110–11.
19 See further Johnson 2013.
20 See for example, Johnson 2015, pp. 83–84.
21 Augoyard and Torgue 2005, pp. 130–31, 136–37.
22 Regarding Shakespeare, see for example, Smith 1999; Johnson 2005; and on film see for example Michel Chion 1994, on acousmatic sound.
23 Zajonc 2004, p. 254.
24 See further LeDoux 1998, especially Chapters 6 (pp. 138–78) and 8 (pp. 224–66).
25 Becker 2001, p. 154.
26 Wardhaugh 2008, p. 105.
27 ibid., p. 59.
28 Clark 2011, p. xxviii.
29 Tribble and Keene 2011, p. 4.
30 Cited Tribble and Keene 2011, p. 3.
31 See for example Goldin-Meadow 2003, pp. 242–44; and McNeill 2005, pp. 147–48, 195–206.
32 Cited in Johnson 2000, p. 181.
33 See for example Grand 2013.
34 Faulkner 2013, pp. 41, 42.

References

Augoyard, J. -F. and Torgue, H., 2005. *Sonic experience: A guide to everyday sounds*, trans. A. McCartney and D. Paquette, Montreal: McGill-Queen's University Press.

Bacon, F., 1960. 'The great instauration, 1620'. In F.H. Anderson, ed., *The new organon and related writings*, Indianapolis, New York, Kansas City, USA: Bobbs-Merrill, pp. 3–29.

Becker, J., 2001. 'Anthropological perspectives on music and emotion'. In J.A. Sloboda and P.N. Juslin, eds., *Music and emotion: Theory and research*, Oxford: Oxford University Press, pp. 135–69.

Chion, M., 1994. *Audio-vision: Sound on screen*, trans. C. Gorbman, New York: Columbia University Press.

Clark, A., 2011. *Supersizing the mind: Embodiment, action, and cognitive extension*, Oxford and New York: Oxford University Press.

Corbin, A., 1998. *Village bells: The culture of the senses in the nineteenth century French countryside*, trans. M. Thom, Columbia: Columbia University Press.

Faulkner, R., 2013. *Icelandic men and me: Sagas of singing, self, and everyday life*, Farnham, Surrey: Ashgate.

Feld, S., 2012. *Sound and sentiment: Birds, weeping, poetics and song in Kaluli expression*, 3rd edn, Durham, NC: Duke University Press.

Goldin-Meadow, S., 2003. *Hearing gesture: How our hands help us think*, Cambridge, MA and London: The Belknap Press of Harvard University Press.

Grand, D., 2013. *Brainspotting: The revolutionary new therapy for rapid and effective change*, Boulder, CO: Sounds True.

Jameson, F., 1981. *The political unconscious: Narrative as a socially symbolic act*, Ithaca, NY: Cornell University Press.

Järviluoma, H., ed., 1994. *Soundscapes: Essays on vroom and moo*, Tampere and Seinäjoki, Finland: Department of Folk Tradition and Institute of Rhythm Music.

Järviluoma, H., Kyto, M., Truax, B., Uimonen, H. and Vikman, N., 2009. *Acoustic environments in change & five village soundscapes*, Tampere: Tampereen ammattikorkeakoulu.

Järviluoma, H. and Wagstaff, G., 2002. *Soundscape studies and their methods*, Helsinki: Finnish Society for Ethnomusicology and Department of Art, Literature and Music.

Johnson, B., 2000. *The inaudible music: Jazz, gender and Australian modernity*, Sydney: Currency Press.

Johnson, B., 2005. 'Hamlet: Voice, music, sound', *Popular Music*, 24/2, pp. 257–67.

Johnson, B., 2013. 'I hear music: Popular music and its mediations', *IASPM@Journal Online*, 3/2, Special issue: 'Popular music studies in the twentieth-first century', pp. 96–110.

Johnson, B., 2015. ' "Lend me your ears": Social policy and the hearing body'. In S. Homan, M. Cloonan and J. Cattermole, eds., *Popular music and cultural policy*, New York and Abingdon, Oxon: Routledge, pp. 79–91.

Johnson, B. and Cloonan, M., 2008. *Dark side of the tune: Popular music and violence*, Aldershot, UK and Burlington, VT: Ashgate.

LeDoux, J., 1998. *The emotional brain*, London: Phoenix.

MacNevin, R., 2005. 'Editorial', *Soundscape: The Journal of Acoustic Ecology*, 1/2 (Winter), p. 4, Accessed 29 July 2015, http://multimedia.3m.com/mws/media/893200O/soundscape-the-journal-of-acoustic-ecology.pdf.

Manning, F., 2012. *The middle parts of fortune*, London: The Folio Society. (First published in 1929, and then in a revised form as *Her privates we*).

McNeill, D., 2005. *Gesture and thought*, Chicago and London: University of Chicago Press.

Picker, J.M., 2003. *Victorian soundscapes*, Oxford: Oxford University Press.

Pinch, T. and Blijsterveld, K., eds., 2011. *The Oxford handbook of sound studies*, Oxford and New York: Oxford University Press.

Russolo, L., 1986. *The art of noises*, trans. B. Brown. New York: Pendragon Press. (First published in 1913).

Schafer, R.M., 1977. *Five village soundscapes*, Tampere: University of Applied Science.

Schafer, R.M., 1977. *The soundscape: Our sonic environment and the tuning of the world*, Toronto: McClelland and Stewart.

Smith, B.R., 1999. *The acoustic world of early modern England: Attending to the O-factor*, Chicago and London: University of Chicago Press.

Sterne, J., 2003. *The audible past: Cultural origins of sound reproduction*, Durham, NC: Duke University Press.

Tribble, E.B. and Keene, N., 2011. *Cognitive ecologies and the history of remembering: Religion, education and memory in early modern England*, Basingstoke: Palgrave Macmillan.

Wardhaugh, B., 2008. *Music, experiment and mathematics in England, 1653–1705*, Farnham and Surrey: Ashgate.

Westerkamp, H. and Truax, B., 1978. *Handbook of acoustic ecology*, Burnaby, BC Canada: ARC Publications.

Wilson, E., 2004. 'Gut feminism', *Differences: A Journal of Feminist Cultural Studies*, 15/3, pp. 66–94.

Zajonc, R.B., 2004. *The selected works of R.B. Zajonc*, New Jersey: Wiley.

Part I

Sound and Voice

2 "The World Wanderings of a Voice"

Exhibiting the Cylinder Phonograph in Australasia[1]

Henry Reese

'The world wanderings of a voice.' It was in such grandiose terms that 'Professor' E. Douglas Archibald described his global travels in the early 1890s. Leaving behind metropolitan Britain in 1890 for what he later termed the 'Greater Britain of the South,' the English scientist-turned-impresario embarked on a brief Antipodean career in sound. Over the following two years, Archibald traversed the theatres and town halls of South and Southeast Asia, Australia and New Zealand, unveiling the contents of his so-called 'box of voices and music' to enthusiastic audiences wherever he went.[2]

This 'box of voices' was the recently completed 'Perfected' Phonograph, a device patented by the Edison Phonograph Company in 1888 in response to the invention of the rival Columbia Graphophone some two years previously. Unlike the prototypic talking machines of the late 1870s, which inscribed sound into fragile strips of tinfoil, the Perfected Phonograph was a strong and portable device. For the first time, it was possible to record sound onto durable wax cylinders, to transport these cylinders around the world and to reproduce the sounds etched thereon multiple times, in what was described as a 'calm defiance of time' by a colleague of Archibald's in 1893.[3]

This chapter offers the first sketch of the route and reception of Douglas Archibald's phonographic tour in Australia and New Zealand. Archibald's two-year traversal of urban and rural auditoriums is important for a number of reasons, as much for his mobility itself as the content of his performances. The progress and internal logic of Archibald's tour speaks to the resonance of staged participatory media, and the robust cultural networks across which such cultural exchange occurred, in a period immediately preceding the commercialization of recorded sound in Australasia. The content and reception of Archibald's demonstrations underline key tensions in the Australasian auditory culture of the period. Tropes of distance and difference were deployed and negotiated in complex ways through performance.

The tour was also important as a sonic phenomenon. Archibald's performances were concerned with specific sounds and acoustic environments as much as with the science of sound reproduction itself. The English meteorologist was an enthusiastic recordist as well as a showman; he recorded voraciously as he travelled throughout South and Southeast Asia, colonial Australia and New Zealand. He integrated his field recordings into his performances, juxtaposing an array of familiar and foreign sound events before the eager ears of his auditors. Douglas Archibald did not merely snatch sound events out of the immersive sonic tapestry through which he travelled, as recordists working within earshot of the World Soundscape Project have striven to do since the mid-twentieth century.[4] Such a disposition towards recording was not technologically possible at the time. Rather, Archibald carefully recorded, selected and curated the sounds of his own global travels, guiding specific sounds into the horn of his phonograph and guiding his audiences towards an understanding of the sounds of the wider world and Empire. This chapter is an attempt to follow these sounds.

The Phonograph Exhibition as a Cultural Form

In order to explore what the British Empire sounded like, it is necessary to open the mobile phonograph out to the wider soundscapes through which it moved, to examine those who took sound seriously and engaged deliberately and knowledgeably with it. To do so is to bring together two avenues of research that often exhibit divergent trajectories. The first of these is historical soundscape studies. Research in this field tends to drift outwards from the perceiving subject, attending to the intimate, emplaced 'wraparound' of sound in a given acoustic community.[5] To 're-compose,' in Maarten Walraven's terms, the 'interplay of humans and their environment through sound' becomes a question of identifying and describing the social role of those sounds that were significant in the everyday lives of their auditors.[6]

This disposition sits in marked contrast to the historiography of sound recording, which tends to move inward, placing the phonograph within a history of mechanical reproduction, audio fidelity and music industry politics.[7] Yet there is also a sound precedent for placing these research endeavours within the same analytic frame. For Daniel Morat, questions of technology and media are the hallmarks of *modern* auditory cultures.[8] Douglas Kahn has argued that the idea and possibility of phonography made it possible to deal with sound as a cultural phenomenon in the late nineteenth century; recording introduced 'the fact of worldly sound into culture,' thereby 'making the boundaries between humans and machines, writing and voice, human sounds and worldly sounds, music and noise, much more problematic.'[9] Here the early phonograph can be understood as part of an emergent modern disposition towards sound and listening; Western publics, Emily Thompson has noted, became 'sound conscious' in a way that recast sound as a general category of acoustic experience.[10]

The cultural context of the early phonograph encouraged listeners to think about sound itself. Turn-of-the-century auditors felt that the phonograph could separate 'the spoken word from the body of the speaker, just as it does the sound of a trumpet from the instrument itself, or street noises from the city.'[11] This chapter argues that phonography's imagined relation to (modern) sound, as what Jonathan Sterne has called an 'object of knowledge' in its own right, mandates closer attention both to early explorations of sound recording and to the wider soundscapes within which these sonic encounters occurred.[12]

Public displays of recorded sound were a key site for such auditory negotiations. Exhibitions constituted a notable, if short-lived, feature of the new inscriptive medium from its inception in 1877, particularly in North America.[13] Tinfoil phonographs were also displayed in Australia as early as 1878, alongside a host of other sensory spectacles, at Melbourne's popular scientific *conversazioni*.[14] The emergence of the wax cylinder phonograph in the late 1880s inspired a second, global wave of sonic display. Patrick Feaster, William Howland Kenney and Annegret Fauser, among others, have offered rich accounts of displays of wax cylinder phonographs in parlours, arcades and exhibitions, emphasizing the wonder and amazement that often crystallized around the fact of sound reproduction in the 1880s and 1890s.[15]

Despite the early industry's globally expansive rhetoric (such as common claims that Edison records 'echo all over the world'), few extant studies have examined the phenomenon of the phonograph exhibitions outside European or North American contexts.[16] This narrative is further complicated by the fact that not all phonograph demonstrations were equal; the form of entertainment that Douglas Archibald exemplified was neither long-lived nor internally stable. Unlike earlier performances, what was on display by the 1890s was no longer exclusively what Gustavus Stadler has called 'the quasi-magical fact of sound reproduction.'[17] The phonograph display was not *just* a 'marvel' or 'modern wonder' in the mode identified by Bernhard Rieger; its content was now also of significance.[18] Does the phonograph of the 1890s sound different in Australasia?

On the Road with Douglas Archibald

Douglas Archibald has been described as the first person to introduce the wax cylinder phonograph to Australia and New Zealand.[19] Hitherto, however, no extensive treatment of his life and career has been written. In Ross Laird's 1999 survey of early sound recording in Australia, Archibald's tour is glossed as part of the 'prehistory' of the Australian recording industry.[20] Where he appears elsewhere in the literature, he is depicted as an ephemeral figure among a wider field of modern entertainments in Australia, chiefly early cinematography.[21]

Two of the most sophisticated treatments of Douglas Archibald do not consider his mobility or biography at large, but rather explore the English lecturer's impact on specific Asian communities. In a recent monograph on European 'everyday technologies' in colonial Sri Lanka, Nira Wickramasinghe describes Archibald's demonstrations as a precursor to the dynamic recording culture that flourished in Ceylon around the turn of the twentieth century.[22] Likewise, Suryadi, a historian of Indonesia, has contextualized the British showman's ill-fated 1892 Javanese tour in terms of the wider networks of European and Anglo-Australian entertainers active in the fin de siècle Dutch East Indies.[23] Both accounts provide synchronic instances of the impact of Archibald's novel spectacle, depicting the popular cultures of colonial South and Southeast Asia as enmeshed between local tradition, colonial influences and modern consumer goods.

Little remains of Archibald's tour today. A crumpled brochure and a faded photograph attest to a spectacle that was once described as the 'most marvellous' ever to have graced the Australasian colonies.[24] For historian Melissa Bellanta, this is a common methodological challenge: despite their cultural impact, 'popular' stage entertainments exhibit a frustrating tendency to slink out of the historical record after the final fall of the curtain.[25] As Richard Waterhouse has influentially noted, this is particularly pronounced in the Australian context, where class divisions were not as exclusive, the theatre not as rigidly 'bifurcated' as on the British or American stages even in the late nineteenth century.[26] The result has been that non-elite entertainments, despite their cross-class appeal, do not as readily furnish source material for the historian of the stage. Archibald's entertainment, which elided the scientific and the spectacular and moved from waxwork parlour to town hall—mechanics institute to elite drawing room—was a similarly multivalent, and largely forgotten, cultural form.

Despite the methodological lacunae outlined above, the Edmund Douglas Archibald who emerges from the documentary record is still a colourful figure. For Archibald was not always a moustachioed phonographic showman. His career was mercurial and restless: a patchwork of science, education and entertainment spooled out across the span of the British Empire. Born in Hampstead in 1850, this clubbable judge's son received his education at St John's College, Oxford. Graduating with his MA in 1879, Archibald resided in India from the late 1870s, teaching mathematics under the auspices of the Bengal Educational Service.[27] Returning to England in 1881, Archibald settled in Tunbridge Wells, working as a tutor and experimenting with kites and weather balloons.[28] A member of the Royal Meteorological Society and the British Association for the Advancement of Science, he regularly attended transatlantic conferences and published widely on meteorological matters.[29]

Archibald was also an inveterate correspondent and traveller, with a keen interest in the affairs of the British Empire, particularly regarding matters of climate, infrastructure and administration.[30] He was a scientist with his head, quite literally, in the clouds; writing on the topic of 'Clouds and

Cloudscapes' in 1894, Archibald alternates between typological detail and wide-eyed rapture at the majesty of a skyline that 'frequently outrivals the Himalaya in grandeur.'[31] A similar sense of picturesque difference (curiously entwined with realpolitik) would also undergird Archibald's 1897 call for the Colonial Office to annex the islands of Hawai'i.[32] In the mid-1880s, Archibald served on the Krakatoa Committee of the Royal Society, tracing the global diffusion of vapour plumes from the Sundanese volcano's sensational 1883 eruption.[33] It is with no small measure of irony that, by the end of the decade, his scientific interests had shifted away from what John Picker has noted was the loudest earthly phenomenon ever recorded in its time, and towards the whispering promises of a novel technology that probed the very boundaries of audibility.[34]

As with the airy medium through which it travelled, Archibald's enthusiasm for sound bridged the romantic and the scientific. His shows were, as the *West Australian* noted in 1891, 'at once a concert and a scientific treat.'[35] His oft-noted ability to simplify matters of acoustics for a popular audience is clearly shown in the short brochure printed to accompany the Australian leg of his tour. Following a facsimile reproduction of Thomas Edison's original announcement of the Perfected Phonograph is a short discourse by Archibald on the 'Principle of the Phonograph.'[36] Here he unpacks the 'wonderful' physics of phonography in effusive terms: 'sound consists of a series of wave-like movements of the air, analogous to the waves of the sea.'[37]

Prior to steaming south across a different kind of wave, Archibald became involved with the Edison Company's British operations in 1888, at a time when the exhibitionary popularization of the recently 'perfected' phonograph began in earnest.[38] Accompanied by Colonel George Gouraud, Edison's flamboyant London agent, Archibald's first public demonstrations of recorded sound took place in the capital in mid-December and continued apace the following year.[39] He enjoined a crowded field of commercial demonstrators that quickly spread across the Continent in this nascent phase of the recording industry.[40]

Speaking to the *West Australian* in 1891, Archibald claimed to have 'introduced [the phonograph] into England, most parts of Scotland, Ireland, Germany, and Switzerland.'[41] By 1890, however, he had 'naturally' set his sights on 'the Greater Britain of the south,' where he believed 'a Phonograph of this civilised and highly organised type had never been heard.'[42] Travelling via Edison's laboratories in New Jersey, Archibald alighted in Sydney in late May. Once ashore, he arranged for a private audience with the Governor of New South Wales. It was here, during his first Australian phonographic display, that Archibald delivered a pre-recorded message from British statesman William Gladstone.[43]

This 'vocal letter' to Lord Carrington was often employed as the finale of Archibald's demonstrations, and was reprinted in the accompanying brochure.[44] Gladstone stated that he was 'alike honored and gratified in being the first person to make a communication through the phonograph to Australasia.' The machine would mark 'a new bond of amity between Australasia and the United Kingdom.'[45] Archibald's advertisements in the local periodical press meaningfully played with this sense of geographical distance and communication, trumpeting, for instance, that 'Gladstone will speak in Launceston tonight.'[46]

Likewise, the phonograph was heralded in the New South Wales border town of Albury as an intimate means of global connection. According to the *Border Post*, the phonograph 'bridges' wide distances,

> and to the wanderer in climes far removed from his domestic circle, it is ready to touch the heart with the voice of the dear ones at home; and is capable of being made a link of sympathy between the English speaking races of the globe that may be at once unique and effective.[47]

Following his initial Sydney reception, Douglas Archibald journeyed onward to Melbourne. Here he began a season of exhibitions under the management of MacMahon Brothers, a theatrical company which later arranged Archibald's provincial travels along the well-worn popular stage circuits of the period. The MacMahons' hyperbolic advertising copy reinforced the shows as a unique, 'startling' experience, in which 'conversations that have taken place in Europe and America . . . will be actually reproduced,' and 'human speech may be preserved for ever.'[48] As Brian Hochman has recently noted, the salvage rhetoric animated by the early recording business was sufficiently broad to unite an array of disparate cultural modes, from ethnography, to writing, to death.[49]

The tour also relied on careful management of the steam and rail networks that conveyed Archibald from stage to stage, as well as of inter-colonial information streams. A systematic advertising campaign and widespread syndication of colonial newspaper content resulted in the cultivation of a sense of anticipation as the charismatic Englishman toured the continent, assuring him, with few exceptions, of uniformly large audiences. Newcastle's *Morning Herald* wrote in late October 1890, for instance, that 'much has been heard in recent years of Edison's wonderful invention . . . and at last the marvellous instrument is to be exhibited in Newcastle.'[50] The Melbourne *Age*, similarly, reported in 1890 that 'great persuasion, expressed in hundreds of letters from all parts of the colony,' constantly inspired the tour's ineluctable onward motion.[51]

After his initial seven-week season in the Melbourne Waxworks, Archibald spent nearly two months touring the town halls of regional Victoria, often with little more than a day's break between towns.[52] Afterwards, he travelled to Newcastle and Sydney, before covering Tasmania and Adelaide by the end of the year. Archibald's next six months were spent criss-crossing New Zealand, where he remained until June 1891. One striking illustration of the popularity of Archibald's tour in New Zealand can be found in the decision by the local authorities of Greymouth, on the South Island, to put on a 'special phonograph train' to convey listeners to his May 1891 exhibition.[53]

Upon returning to Australia in June, Archibald revisited Tasmania before crossing the Great Australian Bight and spending a month in the colony of Western Australia.[54] By the start of 1892, he had left the Antipodes altogether, spending a number of months performing for predominantly European male audiences in a host of clubs in Ceylon, India, Burma, Singapore and Java.[55] Returning to Australia in early August 1892, Archibald completed his circumnavigation of the continent with a final run of dates in North Queensland and Brisbane.[56] By late September, he had recrossed New Zealand and steamed for America to perform at Chicago's Columbian Exposition of 1893.[57] This departure brought an end both to Archibald's tour and to his career as a phonographic showman. For the following 20 years, he resumed an interest in meteorology and droughts, eventually taking up a chair at Calcutta University.[58]

On Stage with Douglas Archibald

Archibald's exhibitions largely took place in large concert halls or theatres. Admittance costs ranged from one to three shillings, resulting in a socially stratified but not exclusively elite attendance. During a time of economic depression and increasing social disquiet, however, such fees ensured that Archibald's entertainment remained beyond the means of many, and a discount in Melbourne was interpreted as 'conferring a favour on the community.'[59] An Adelaide correspondent writing under the pseudonym of 'Disgusted' nevertheless found such charges to be not only expensive, but also humiliating for status-conscious audience members.[60]

Those that could afford to attend Archibald's performances witnessed an exhibition that consisted of a number of parts. First came an 'intensely interesting' opening lecture and magic lantern show in which the lecturer briefly adumbrated the current state of the science of acoustics and the workings of the phonograph, making explicit an imagined equivalence of human ear and mechanical talking machine.[61] This was followed by a demonstration of the device that took the form of a variety show of sorts, something of an auditory cross-section of the British and American popular stage repertoire of the early 1890s.[62] Speeches, operatic and comic songs, instrumental solos, minstrel songs and brass band numbers were all 'reproduced' live. Audience participation was encouraged, and Archibald often made recordings onstage, involving local musicians, politicians and community figures.

Archibald's collection of famous voices was not limited to that of William Gladstone. Tasmanian Governor Robert Hamilton dictated (or rather shouted, as the technology demanded) a 'Dream of Federation' speech into Archibald's phonograph in 1890, in which he envisioned a future federation of 'all English speaking peoples.'[63] Other voices included Sir George Grey, Western Australian Premier John Forrest, the Anglican Bishop of Colombo and the Maharajah of Gwalior.[64] Many of these speeches were not merely trivial epigrams, but themselves evince an awareness of the planned onward movement of the tour itself, as well as a confidence that these waxen epistles would be heard, and valued, by their recipients; Colin Christie, the Mayor of Newcastle, thrilled in 1890 at the opportunity to address the Mayor of Melbourne directly 'through the medium of this wonderful machine.'[65]

Reviews of Archibald's exhibitions assume a homogeneous listening experience. Reports often claimed, for instance, that the phonograph was 'heard clearly from end to end of the building,' or 'all over the world.'[66] The Melbourne paper *Table Talk* asserted that everything could be heard clearly without any undue 'strain upon the mind.'[67] 'The sounds' of a recorded ballad, too, were 'startlingly real. The song a man had sung into the instrument a couple of years ago came upon the ear as full and clear as if he were singing in some other room.'[68]

Reviews also make continual reference to the distance of the recording site, as well as the numerous times each recording had previously been played: '[s]ome of the phonograms,' reported the *Sydney Morning Herald* in 1890, 'have been repeated eight or nine hundred times without appreciably deteriorating.'[69] This rhetorical insistence on the spectacle of local communities assembling to listen together to the staged sounds of difference and distance more closely resembles more dispersed modes of technological dissemination, such as the earlier 'telephone concert' (exemplified by Budapest's famous Telefon Hirmondó) or the later radio broadcast, thereby serving to underscore the fruitful interplay between cultural forms in this period of media plasticity.[70] A similar conjunction can be found in Archibald's decision, in July 1890, to connect his phonograph to the Melbourne telephone exchange in order to 'guide' a number of records down the wire to the regional Victorian town of Geelong.[71]

For the men and women in Archibald's fashionable audiences, their very first encounter with recorded sound often came in the form of the rich strains of a cornet solo, a polka, played by Arthur Smith of London's Covent Garden Company. The *Ballarat Courier* recounted the experience of these vibrations from the metropole in vivid terms, noting that Smith's record initially produced a 'deep' and portentous silence, whose tensions finally gave way to 'tumultuous applause.'[72] This pattern of dumbstruck wonder slowly resolving into wild enthusiasm was repeated time and again across the colonies. Despite being aware of the phonograph's function, the spectacle of acousmatic sound remained deeply powerful for colonial audiences. Moreover, by 1891, news of the cornet player's death had reached the colonies, causing what the *Launceston Examiner* described as a 'melancholy interest to attach to' the familiar solo.[73]

The Soundscapes of the Phonograph

Archibald's tour offered more than what was described in the Victorian goldfield town of Castlemaine as 'songs sung by men years ago in other lands.'[74] A continually notable feature of Archibald's travels was his cumulative, mimetic re-presentation of the keynotes of the colonial soundscape. In addition to musical and oratorical performances, Archibald recorded local sounds in situ, replaying them on stage. Such cylinders reportedly offered 'a realism most striking in its correctness' and made a profound impact on their auditors.[75]

The exhibition programme also changed and grew as the tour progressed, eventually including field recordings taken on many legs of the tour. Even on return dates, audiences were earwitnesses to the sonic evidence of Archibald's 'world wanderings.' For New Zealand's *Poverty Bay Herald*, which described Archibald as 'Mr. Edison's world pioneer,' it was the travelling nature of the tour itself that made possible Archibald's capture of 'every sort of music and utterance which prevail over the world.' Such a collection, which included 'a scene from an American ballroom . . . [t]he chimes of Westminster Abbey, the songs of our Hindu and Bengali cousins, and speeches by leading statesmen all over the world,' would, the paper argued, be simply 'impossible to duplicate or acquire except by undertaking such a tour as Professor Archibald has accomplished.'[76]

In this sense, Archibald's phonograph provoked a deeper engagement with soundscapes, both familiar and foreign. His numerous field recordings were charged with local meaning. Whether it was the so-called 'discordant bray' of Adelaide street musicians,[77] snatches of itinerant Salvation Army bands from urban Melbourne and Christchurch,[78] 'Nautch' songs from northern India,[79] the chimes of the bells of Auckland's Trinity Cathedral[80] or the 'most comical' roar of the bellman of the Western Australian port town of Fremantle,[81] Archibald's technique of framing the socially significant sounds of his immediate urban environment as staged, performative phenomena heightened the awe provoked by the medium.

This suggests two curious listening experiences. Firstly, Archibald's deliberate selection and presentation of unique recordings and valuable speeches—which smacked of scarcity at a time when the mass-production of cylinders was not yet possible—helped construct acoustic knowledge of the wide spaces of other colonial locales, connecting each soundscape by the internal, self-referential logic of the very tour itself. Listeners were invited to engage in an imaginary process of habitation of the familiar and unfamiliar soundscapes reframed and presented before them, which relied on a common aurality and set of sonic topoi for their felicity; what they heard was not just a series of sound recordings, but a carefully curated, recognizable 'acoustic journey.'[82] Secondly, Archibald's framing of local sounds also drew attention to them *as sound as such*, providing a vivid entry-point to the time and place of the records' very creation, and offering a common contextual frame in which his 'world wanderings' were to be understood.

Conclusion

Curled within the spiral tracks of Douglas Archibald's wax cylinders is the promise of recovering everyday negotiations of the soundscapes of a changing Australasia. Archibald's tour, in its mobility, form and reception, underlines not only the construction and vernacular negotiation of what Jonathan Sterne has called 'sociotemporal' and 'sociospatial' networks of sound, but also the methodological value of bringing the concept of the soundscape into conversation with questions of recording and mechanical reproduction.[83] For the first time, the phonograph was capable of evoking the soundscapes of locales both near and far. Through performance, Archibald imaginatively brought the significant sounds of the world alive in his auditors' ears, linking local to global vibrations in one recognizable field of mediated auditory experience.

Douglas Archibald's phonographic travels did not merely stimulate excitement in his colonial auditors; the mobility of the sounds on display also reveals a hitherto under-examined aspect of the history of sound. In the recognizability of Archibald's staged field recordings, and the acoustic environments for which they stood, we can perceive the emergence of an imagined (acoustic) community of Empire as well as a thicker, proto-nationalistic conception of the local in late colonial Australia. As heard, for instance, in Governor Hamilton's Tasmanian 'Dream of Federation' speech, the introduction of the cylinder phonograph to colonial audiences coincided with a period that witnessed the growth of both colonial nationalism and a domestic Federation movement, which would culminate in the creation of the Commonwealth of Australia by 1901. By linking the phonograph to the soundscape, this reconstruction of Douglas Archibald's phonographic travels complements, and complicates, the growing body of scholarship that emphasizes the contribution of soundscapes to the construction and maintenance of the cultural bonds of nation and Empire, across dispersed geographical space. If the sonic environment can, as ethnomusicologist Steven Feld has argued, engender a 'sensual poesis of place,' then Archibald's tour suggests that this sonic 'emplacement' may be technologically mediated, and not limited to any one place or time.[84]

Douglas Archibald's phonograph tour also speaks to connections between histories of specific sound media. Arguably, the directed, evaluative techniques adopted by Archibald's listeners resemble the distant listening cultures of early radio, some three decades later. Archibald's movement is not unlike later broadcasting, although his analogue talking machine lacked the instantaneity and simultaneity of the airwaves. In addition, then, to fleshing out the wider soundscapes of the phonograph, this case study complicates and lengthens the cultural history of public listening in the modern West.[85] Like the wafting 'cloudscapes' of his meteorological imaginings, Archibald's 'world wanderings' were ephemeral and short-lived, scientific and romantic, but ultimately resonant with both local and global meaning.

Notes

1 I would like to express my thanks to Professor David Goodman and Dr Julie Fedor for their invaluable comments on earlier drafts of this chapter. Many thanks are also due to Professors Joy Damousi and Paula Hamilton for convening the wonderful conference from which this chapter originated.
2 Archibald 1893.
3 Bleyer 1893, p. 15.
4 Lane and Carlyle 2013.
5 Feld 1996, p. 100.
6 Walraven 2013. See also Smith 2001, pp. 264–65.
7 Picker 2003, p. 112.
8 Morat 2014, pp. 4–7.
9 Kahn 1992, pp. 5–6, 15.
10 Connor 1997, p. 216; Thompson 2002, p. 59.
11 Gauß 2014, p. 74.
12 Sterne 2003, p. 23.
13 Gitelman 2012, pp. 293–94.
14 Hartrick 2008, p. 11.7.
15 Kenney 1999, pp. 23–43; Feaster 2001; Fauser 2005, pp. 279–311.
16 Gitelman 1999, p. 164.
17 Stadler 2010, p. 87.
18 Rieger 2005, pp. 20–50.
19 'An interview with Professor Archibald', *West Australian*, 22 August 1891.
20 Laird 1999, pp. 2–15.
21 Long 1993, pp. 37–38. See also Brisbane 1991.
22 Wickramasinghe 2014, pp. 83–85.
23 Archibald's Javanese tour was a disaster. A fistfight with a Dutch journalist over allegations of fraud led to a three-week spell in a Surabaya prison in June 1892. Following this altercation, Archibald beat a hasty return to Queensland. See Suryadi 2006, pp. 271–91.
24 'The phonograph', *Mount Alexander Mail*, 1 October 1890.
25 Bellanta 2012b, p. 415.
26 Waterhouse 1995, pp. 84–86.
27 Foster 1888, p. 29; 'Woolwich and Sandhurst entrance examinations', *Morning Post*, 14 August 1885, p. 1.
28 'The Aeronautical Society', *Morning Post*, 6 July 1886, p. 6; 'Kite-flying as a science', *Pall Mall Gazette*, 11 July 1888.
29 'The British Association. Uses of balloons and kites in geographical research', *Freeman's Journal and Daily Commercial Advertiser*, 15 September 1898.
30 See Archibald 1896.
31 Archibald 1894, p. 572. See also Archibald 1897a.
32 Archibald 1897b.
33 Russell and Archibald 1888.
34 Picker 2003, p. 4.
35 'The phonograph was again exhibited to a crowded house at the St. George's Hall last night', *West Australian*, 20 August 1891.

36 Edison 1890.
37 Archibald 1890, 'The principle of the phonograph', p. 13.
38 Gelatt 1977, pp. 100–01.
39 'Edison's perfected phonograph', *Morning Post*, 17 December 1888; 'Olympia. Boxing-day. The phonograph', *Standard*, 25 December 1888, p. 1.
40 See Lynd 1893.
41 'An interview with Professor Archibald'.
42 Archibald 1893.
43 'The phonograph', *Sydney Morning Herald*, 18 October 1890.
44 'Review: The phonograph', *Maryborough and Dunolly Advertiser*, 22 September 1890.
45 Archibald 1890, 'Copy of the phonogram spoken in London on March 8th, 1890', p. 16.
46 'Academy of Music. Monday and Tuesday next, the phonograph', *Launceston Examiner*, 20 June 1891.
47 'Edison's perfected phonograph', *Albury Border Post and Wodonga Advertiser*, 10 October 1890, p. 11.
48 'Athenæum Hall. The phonograph a sensational success', *Age*, 28 June 1890.
49 Hochman 2014, pp. 73–113.
50 'Review: Victoria Theatre. Edison's wonderful phonograph', *Newcastle Morning Herald and Miners' Advocate*, 25 October 1890.
51 'St. George's Hall. Announcement extraordinary! The phonograph', *Age*, 1 November 1890. See also Putnis 2010, pp. 165–68.
52 'Academy of Music. To-night. The phonograph. Edison's startling talking machine', *Ballarat Courier*, 22 September 1890.
53 'The phonograph', *Grey River Argus*, 16 May 1891.
54 'Entertainments: Oddfellows' Hall, Fremantle. The marvel. The talking machine. To-night', *West Australian*, 14 August 1891.
55 'Evening party at Tata House', *Times of India*, 21 March 1892, p. 5. See also Sinha 2005, pp. 185–86.
56 'Townsville, August 16', *Brisbane Courier*, 17 August 1892, p. 5.
57 'Edison's phonograph', *Wairarapa Daily Times*, 12 December 1892, p. 2.
58 'Advertisements', *King Country Chronicle*, 15 February 1913, p. 3.
59 'The waxworks', *Table Talk*, 1 August 1890. See also Bellanta 2012a, pp. 145–46.
60 'Lectures on the phonograph', *Advertiser* (Adelaide), 26 November 1890.
61 'Amusements. The phonograph', *Sydney Morning Herald*, 14 October 1890.
62 Waterhouse 1990, pp. 108–12.
63 'Current topics', *Launceston Examiner*, 22 June 1891.
64 'Theatre Royal. Announcement extraordinary', *Brisbane Courier*, 8 September 1892, p. 2.
65 'Victoria Theatre. The phonograph', *Newcastle Morning Herald and Miners' Advocate*, 29 October 1890.
66 'Edison's wonders. The phonograph', *Newcastle Morning Herald and Miners' Advocate*, 28 October 1890; 'Theatre Royal', *Brisbane Courier*, 6 September 1892, p. 2.
67 'The waxworks', *Table Talk*, 18 July 1890.
68 'The phonograph', *Grey River Argus*, 16 May 1891.
69 'The phonograph', *Sydney Morning Herald*, 18 October 1890.
70 Marvin 1988, pp. 222–31.
71 'The waxworks', *Table Talk*, 18 July 1890.
72 'Academy of Music. Edison's astounding talking machine. The phonograph', *Ballarat Courier*, 24 September 1890.
73 'Academy of Music. The phonograph', *Launceston Examiner*, 23 June 1891.
74 'The phonograph', *Mount Alexander Mail*, 1 October 1890.

75 'Academy of Music. The phonograph', *Launceston Examiner*, 23 June 1891.
76 'The new loud phonograph', *Poverty Bay Herald*, 28 November 1892, p. 2.
77 'The phonograph. Opening night', *Lyttelton Times*, 13 January 1891.
78 'Town Hall. The phonograph, Edison's astounding talking machine, Wednesday and Thursday next, June 17th and 18th. Two nights only', *Mercury*, 12 June 1891.
79 'Edison's loud phonograph', *Wanganui Herald*, 11 October 1892, p. 2.
80 'Review: The phonograph', *Maryborough and Dunolly Advertiser*, 22 September 1890.
81 'The phonograph at St. George's Hall', *West Australian*, 18 August 1891.
82 See Collins 2006.
83 Sterne 2003, pp. 308–09.
84 Feld 1996, pp. 101–05.
85 Lacey 2013, p. 11.

References

Archibald, E. D., ed., 1890. *The perfected phonograph, described by its inventor, Thomas Alva Edison: With an appendix by Professor Douglas Archibald, M.A., Oxon.*, Melbourne: Wm. Marshall & Co.

Archibald, D., 1893. 'The world wanderings of a voice', *The Phonogram: A Monthly Journal Devoted to the Science of Sound and Recording of Speech* 1(2) (June), pp. 22–23.

Archibald, D., 1894. 'Clouds and cloudscapes', *The English Illustrated Magazine* 126 (March), pp. 571–75.

Archibald, D., 1896. *The climate of the hill sanitaria of Ceylon, Kandy, Hatton, Nuwara Eliya, Bandarawela*, Colombo: Observer Printing Works.

Archibald, D., 1897a. *The story of the earth's atmosphere*, New York: D. Appleton and Company.

Archibald, D., 1897b. 'Why not annex Hawaii?' *The English Illustrated Magazine* 163 (April), pp. 93–96.

Bellanta, M., 2012a. 'Naughty and gay? Rethinking the nineties in the Australian colonies', *History Australia* 9(1) (January), pp. 136–54.

Bellanta, M.J., 2012b. 'Australian masculinities and popular song: The songs of sentimental blokes 1900–1930s', *Australian Historical Studies* 43(3), pp. 412–28.

Bleyer, J.M., 1893. 'Living autograms', *The Phonogram: A Monthly Journal Devoted to the Science of Sound and Recording of Speech* 1(1) (May), pp. 14–16.

Brisbane, K., ed., 1991. 'Demonstrations of the phonograph', in *Entertaining Australia: The performing arts as cultural history*, Sydney: Currency Press, p. 114.

Collins, D., 2006. 'Acoustic journeys: Exploration and the search for an aural history of Australia', *Australian Historical Studies* 37(128), pp. 1–17.

Connor, Steven, 1997. 'The modern auditory I', in R. Porter, ed., *Rewriting the self: Histories from the Renaissance to the present*, London and New York: Routledge, pp. 203–23.

Edison, T.A., 1890. 'The perfected phonograph', in *The perfected phonograph, described by its inventor, Thomas Alva Edison: With an appendix by Professor Douglas Archibald, M.A., Oxon.*, Melbourne: Wm. Marshall & Co., pp. 4–12.

Fauser, A., 2005. *Musical encounters at the 1889 Paris World's Fair*, Rochester, NY: University of Rochester Press.

Feaster, P., 2001. 'Framing the mechanical voice: Generic conventions of early phonograph recording', *Folklore Forum* 32(1), pp. 57–102.

Feld, S., 1996. 'Waterfalls of song: An acoustemology of place resounding in Bosavi, Papua New Guinea', in S. Feld and K.H. Basso, eds., *Senses of place*, Santa Fe: School of American Research Press, pp. 91–135.

Foster, J., 1888. *Alumni Oxonienses: The members of the University of Oxford, 1715–1886: Their parentage, birthplace, and year of birth, with a record of their degrees*, vol. I, Oxford and London: Parker & Co.

Gauß, S., 2014. 'Listening to the horn: On the cultural history of the phonograph and the gramophone', in D. Morat, ed., *Sounds of modern history: Auditory cultures in 19th- and 20th-century Europe*, New York and Oxford: Berghahn, pp. 71–100.

Gelatt, R., 1977. *The fabulous phonograph: 1877–1977*, 2nd revised edn, New York and London: MacMillan.

Gitelman, L., 1999. *Scripts, grooves, and writing machines: Representing technology in the Edison era*, Redwood City, CA: Stanford University Press.

Gitelman, L., 2012. 'The phonograph's new media publics', in J. Sterne, ed., *The sound studies reader*, London and New York: Routledge, pp. 283–303.

Hartrick, E., 2008. ' "Curiosities and rare scientific instruments": Colonial conversazioni in Australia and New Zealand in the 1870s and 1880s', in K. Darian-Smith, R. Gillespie, C. Jordan and E. Willis, eds., *Seize the day: Exhibitions, Australia and the world*, Melbourne: Monash University ePress, pp. 11.1–11.19.

Hochman, B., 2014. *Savage preservation: The ethnographic origins of modern media technology*, Minneapolis and London: University of Minnesota Press, 2014.

Kahn, D., 1992. 'Introduction: Histories of sound once removed', in D. Kahn and G. Whitehead, eds., *Wireless imagination: Sound, radio, and the avant-garde*, Cambridge, MA and London: MIT Press, pp. 1–29.

Kenney, W. H., 1999. *Recorded music in American life: The phonograph and popular memory, 1890–1945*, New York and Oxford: Oxford University Press.

Lacey, K., 2013. *Listening publics: The politics and experience of listening in the media age*, Cambridge and Malden, MA: Polity.

Laird, R., 1999. *Sound beginnings: The early record industry in Australia*, Sydney: Currency Press.

Lane, C. and Carlyle, A., 2013. 'Introduction', in C. Lane and A. Carlyle, eds., *In the field: The art of field recording*, Axminster: Uniformbooks, pp. 9–13.

Long, C., 1993. 'Australia's first films: Facts and fables—Part one: The kinetoscope in Australia', *Cinema Papers* 91 (January), pp. 36–43.

Lynd, W., 1893. 'On tour with the phonograph', *The Phonogram: A Monthly Journal Devoted to the Science of Sound and Recording of Speech* 1(1) (May), pp. 7–8.

Marvin, C., 1988. *When old technologies were new: Thinking about electric communication in the late Nineteenth century*, New York and Oxford: Oxford University Press.

Morat, D., 2014. 'Introduction', in D. Morat, ed., *Sounds of modern history: Auditory cultures in 19th- and 20th-century Europe*, New York and Oxford: Berghahn, pp. 1–12.

Picker, J. M., 2003. *Victorian soundscapes*, Oxford: Oxford University Press.

Putnis, P., 2010. 'News, time and imagined community in colonial Australia', *Media History* 16(2), pp. 153–70.

Rieger, B., 2005. *Technology and the culture of modernity in Britain and Germany, 1890–1945*, Cambridge and New York: Cambridge University Press.

Russell, F.A.R. and Archibald, E. D., 1888. 'Part IV: On the unusual optical phenomena of the atmosphere, 1883–6, including twilight effects, coronal appearances, sky haze, coloured suns, moons, &c', in G. J. Symons, ed., *The eruption of Krakatoa and subsequent phenomena: Report of the Krakatoa committee of the royal society*, London: Trübner & Co, pp. 151–463.

Sinha, M., 2005. 'Britishness, clubbability, and the colonial public sphere', in T. Ballantyne and A. Burton, eds., *Bodies in contact: Rethinking colonial encounters in world history*, Durham and London: Duke University Press, pp. 183–200.

Smith, M.M., 2001. 'Sound matters: An essay on method', in *Listening to Nineteenth-century America*, Chapel Hill and London: University of North Carolina Press, pp. 261–69.

Stadler, G., 2010. 'Never heard such a thing: Lynching and phonographic modernity', *Social Text* 28(1) (Spring), pp. 87–105.

Sterne, J., 2003. *The audible past: Cultural origins of sound reproduction*, Durham and London: Duke University Press.

Suryadi, 2006. 'The "talking machine" comes to the Dutch East Indies: The arrival of Western media technology in Southeast Asia', *Bijdragen Tot de Taal-, Land- En Volkenkunde* 162(2/3), pp. 269–305.

Thompson, E., 2002. *The soundscape of modernity: Architectural acoustics and the culture of listening in America, 1900–1933*, Cambridge, MA and London: MIT Press.

Walraven, M., 2013. 'History and its acoustic context: Silence, resonance, echo and where to find them in the archive', *Journal of Sonic Studies* 4(1) (May), Available at http://journal.sonicstudies.org/vol04/nr01/a07.

Waterhouse, R., 1990. *From minstrel show to vaudeville: The Australian popular stage 1788–1914*, Sydney: New South Wales University Press.

Waterhouse, R., 1995. *Private pleasures, public leisure: A history of Australian popular culture since 1788*, Melbourne: Longman.

Wickramasinghe, N., 2014. *Metallic modern: Everyday machines in Colonial Sri Lanka*, New York and Oxford: Berghahn.

3 "Are You Sitting Comfortably?"

The Changing Position of Storytellers on Early Australian Radio

Jennifer Bowen

Reading-aloud and listening-to-reading-aloud of books have a lengthy tradition in the West: in the ancient world, they were a mode of reception for both sacred texts and the latest writings of poets and scholars.[1] In the nineteenth and early twentieth centuries, listening to the words of the Bible or a work of fiction was a household pastime in many parts of the English-speaking world, alongside the sharing of family letters as migration rose.[2] Authors became celebrated as readers of their own works, from the large public audiences of Charles Dickens to the more select gatherings of the 'Inklings.'[3] Whether for pleasure or improvement in this life or next, these were activities at which reader and listeners were together in time and place to the point where the shared character of the occasion could outweigh the significance of the text, which might already be familiar to listeners. The invention of sound recording enabled the radical separation of sound and listening, and Thomas Edison enthusiastically predicted a future of recorded books, extolling the 'profit and amusement for the lady and gentleman whose eyes and hands may be otherwise employed' as well as 'the greater enjoyment to be had from a book when read by an elocutionist.'[4] The first 'talking books' in the 1930s were cumbersome affairs, and it took the development of audiocassettes in the 1970s for the production of audiobooks to acquire a commercial footing. The rise of digital technology and podcasting has led to an enormous growth in audiobooks, which in turn has spurred the development of related scholarly interest.[5]

Book readings on the radio are, however, a different matter. Audiobook users share much of the autonomy of a reader of print material: choice about the text to listen to as well as when and where to listen. The radio listener has no opportunity to pause, rewind and re-hear. The serialized reading of books over the air was part of the output of many radio stations when broadcasting began and can still be heard in the twenty-first century in various parts of the world.[6] Radio histories have paid scant attention to the role of book readings, despite a growing number of investigations into the interactions of writers and early broadcasting.[7] Neither has the discussion of the literary encounter to which audiobooks give rise been accompanied by analyses of radio book readings in the past or present.[8] Yet book reading on radio has an important history, in which broadcasters' choices in relation to readers, books and scheduling have a complex relation to listeners. In the Australian instance, the *sound* of the voice was at the heart of an encounter between broadcaster and audience that was not the same from its early years to later practice.

In this chapter I will explore the changing face of book reading on Australian radio.[9] Book readings were widespread before World War II, with many radio stations boasting their own 'storyteller': invariably male, he was one of the stars of the station and promoted as such. After the war, women entered the ranks of readers, only for attention to shift from the reader to the book. Tracking these changes reveals further aspects of the developing role of the broadcast voice in radio listeners' lives. It throws light on how the radio audience was being initiated into the complexity of mass communication in the modern world—the shifting juxtaposition of the impersonal and the personal, the mass and the individual, the solitariness and the sharing—all of which played a pivotal role in the condition of modern life.

The 'Official Fictioneers'

Radio's potential as a vehicle for books was presented in a 1923 press report predicting a bright future for broadcast readings and addresses by authors.[10] Indeed, books were to play a major part on emerging radio as the source of dramas and the subject of talks, whether reviews of new fiction or 'lecturettes' on the classics.[11] The broadcast of readings began with stories for children, who were targeted as listeners from the start of radio services in Australia.[12] *The Listener In*, one of several listings magazines, was launched with its front cover showing a small girl smiling in her cot under a pair of headphones;[13] the caption reminded the public that she was Pat Wilson, the '1924 Empire Baby Champion.'[14] The toddler from East Melbourne had been judged winner in a field of 60,000 entrants, and it was a canny move by *The Listener In* to gain her endorsement of wireless. The reading of books over the air became a staple of children's programmes: 'Uncle Jim' of Brisbane radio station 4GQ wrote in a 1926 'letter to radio pals' that 'the Sandman' was about to complete 12 months of bedtime storytelling.[15] At a radio exhibition in Brisbane in 1928, live performances by bedtime storytellers drew an audience of 32,000 people over 12 nights.[16] While men constituted the majority of children's storytellers, there were women too: 4GQ's 'Little Miss Brisbane,' otherwise the Queensland Radio Service's director's secretary, Mavis MacFarlane, received mail from across Australia and New Zealand in response to her bedtime story broadcasts.[17] In 1928, the children's book reviewer for the *Sydney Mail*, Ella McFadyen ('Cinderella'), discussed *More Stories and How to Tell Them*, a new storybook that included 'a few helpful hints on how to re-tell these narratives to best effect.'[18] McFadyen made a point of reassuring her readers that although 'wireless uncles are, no doubt, a great idea,' the absence of any opportunity for the listener to interrupt or interact means 'the personally-told story will always hold its place.'[19]

By the end of the 1920s, radio programme listings began to include a 'storyteller' at times outside the children's programmes.[20] The existing evidence is not conclusive on whether storytelling programmes intended for a general audience were broadcast earlier in the decade; that is, if they began around the same time as children's storytelling slots, or if they developed after those for children had been running for some years, as a spin-off from them. Certainly they became sufficiently well established in the early 1930s for the *Sydney Morning Herald* to report a speech by the Australian writer Dale Collins at the Millions Club in 1934 predicting the end of reading due to the activity of radio:

> Mr Collins said that the radio storyteller was replacing the novelist . . . the novel reader would become a radio listener; the novel writer would have to cultivate a pleasant speaking voice. . . . 'I do think that the day is coming when the bad writer with a good voice will be preferred to the good writer with a bad voice,' he said. 'Wireless, of course, is to blame—or is it to be praised—for such a prospect.'[21]

Collins implied that there was a threat to novelists from radio storytellers, and sometimes the storytellers did speak from their own scripts.[22] However, they were predominantly reading books written by others: Collins' own *Jungle Maid* was serialized by Melbourne station 3LO in 1934. Debra Adelaide has written about the significant benefit to Australian writers in the 1930s of fees for radio serializations, and this included readings as well as dramatizations.[23] *Wireless Weekly* magazine for January 1935 lists story sessions several times a week on commercial radio stations 2SM, 2GB, 2CH, 2UE and 2UW in Sydney as well as the publicly funded ABC station, 2FC.[24] Author Grafton Burnard took up his pen to the *Sydney Morning Herald* in response to Dale Collins:

> In Australia, the radio has tended to make people very book-minded. . . . I know of very many cases of people who before the introduction of radio rarely read a book of any kind. But now the radio story-teller, with his [sic] short story and his serial, has quickened their interest in literature. . . . Readers (or listeners, as Mr Collins will have them) will not wait to hear just a scrap of their story broadcast, frequently at an inconvenient hour; they will buy the book.[25]

Burnard's letter underscores the benefit to writers of radio's engagement with reading as a promotion for the sale of books. This contrasts with the large-scale newspaper serialization of novels that preceded publication; radio serializations took place after the book was published and available for purchase.

The 1930s saw a growth in radio storytelling across the country, as regional stations became associated with their own storyteller and metropolitan stations began to specialize.[26] In February 1934, Radio 2BL announced it had four 'official fictioneers': Edward Howell for historical novels, plus detective and fantastic stories, Victor Gouriet for humorous and short stories, 'Elizabeth' for a morning serial and Tal Ordell for Australian writing.[27] Radio 2UE billed Ronald Morse for detective fiction and Si Meredith as a general reader.[28] In reporting on the rise of radio storytelling later that year, *The Bulletin* magazine speculated as to whether it would come to be recognized as an art:

> It is already probably the most widely popular feature (outside the cricket scores) of the daily fare provided by the city stations. Most widely popular because most people who have 'no ear for music' like stories. . . . Perhaps in a few years' time it will stand so high that it will command the respect accorded to a Paderewski or a Melba.[29]

The Voice of the Storyteller

The performance of high-profile radio storytellers attracted considerable attention in the broadcast magazines and general press; in fact, their ability to use their voice received more comment than the books they were able to bring to life with it. One of the first to emerge was British-born Captain A.C.C. Stevens, who broadcast in Sydney on 2BL and later 2UE.[30] He was a former soldier and well known as the organizer for the 1926 Sydney Cancer Research Appeal; on radio, he broadcast a daily morning serial using published books, short stories and manuscripts, as well as a weekly motoring talk. There are few details of his selections, other than a run of short stories in April/May 1935 which featured work by older British writers.[31] In April 1935, listeners were invited to send in stories for a competition for which the prize was one guinea 'for every story accepted for telling by Captain Stevens in his own inimitable manner.'[32] There is little detail about that manner, and a profile about Captain Stevens in 1934 begins with a description of his overall physical appearance before characterizing his voice as 'rich, cultured and mellow.'[33]

> A most remarkable gift is the Storyteller's ability to change his [sic] voice to order. . . . By speaking in half a dozen different tones he can put his stories over as conversation among his characters.[34]

This suggests that Stevens' session was almost a drama production, and one report reveals his adeptness with sound effects.[35] Fully cast and produced drama was well established on radio by this time, with nightly serials and plays; however, promotional write-ups ensured that there was no misunderstanding that Stevens was a single speaker. Furthermore, he himself chose the material he read on air.[36] Likewise the published photographs of other storytellers enabled listeners to put a face to the voice, and it was frequently reported that the books serialized were their own selection. This had the effect of personalizing the reading, so that even before a sound was uttered it had 'the trace of the storyteller,' as Walter Benjamin termed it.[37] Familiarity with the voice of the reader—Barthes' 'grain of the voice'—had in many cases been laid down over years.[38] Several storytellers had previously appeared in the children's programmes reading bedtime stories, a particular ritual in the everyday life of a child.[39] More than the published reproduction of a photograph, the memory of a voice is both intimate and personal. The sound of the storyteller can be knitted into the listener's own biography.

While Sydney's Captain Stevens was known for his fundraising work as well as his numerous broadcasts, the Melbourne reader 'Scribe' was also known to the public from work outside radio as well as on air: he had organized a veteran's charity for TB sufferers and regularly gave commentaries for Anzac Day, an annual ritual within Australian life.[40] 'Scribe' is probably the best documented of the radio storytellers, not only because he broadcast for the Australian Broadcasting Commission, whose written archive is voluminous, but also as he had a substantial correspondence with Australian writer Arthur Upfield.[41] Details are available both of the books he read on air and the correspondence he received from listeners, as well as internal ABC deliberations about his broadcasts. 'Scribe' read books aloud over ABC state, then national, radio at 9.35 every morning from 1931 until January 1943, with short breaks in 1940 and 1941.[42] The use of a performance name was a feature of ABC broadcasts at the time, with political commentary given by 'the Watchman' and household hints from 'Domus.' The name 'Scribe' calls to mind that a scribe is one who copies out texts, and here the task was to copy into an oral/aural form. There was no mystery about Scribe's identity; he was Leslie Williams, an actor.[43]

His popularity as a radio reader is demonstrated convincingly in a prolonged internal correspondence between ABC Federal and Victorian managements in 1938, when the Federal Controller of Programmes proposed that the ABC morning serial should be relayed from a single source rather than several state broadcasts.[44] The Manager for Victoria urged that Scribe, the Victorian reader, should handle this session because 'if he is displaced from our programmes, we feel sure that a very difficult situation would arise [as] he has built up a very big following—his mail is the largest of any artist.'[45] The federal plan was delayed and only took place later, with Scribe as the national reader. Scribe's photograph appeared frequently, and in 1937 the ABC sent him, accompanied by Upfield, on a 'boosting tour' to northern Victoria.[46] The books Scribe read were mostly of his own choosing and had wide appeal; they were almost exclusively written by Australian authors and generally set in regional locations. They took in adventure stories, mysteries and some romance—Scribe included the literary Vance Palmer under suffer-ance and Arthur Upfield from choice.[47] Upfield was an increasingly popular writer in the 1930s, but his exclusion from serious consideration by liter-ary critic Nettie Palmer fuelled a life-long loathing by Upfield, which was extended equally to her husband, the writer Vance Palmer.[48] This animosity became the subject of many of the letters between Upfield and Scribe, as Scribe cheerfully recounted the greater number of listeners' letters he had received in response to readings of Upfield's books as against any by Palmer or his friends.[49] The extent of these indicates the desire of listeners to articu-late and communicate their responses to the readings. In addition, some would ask how to obtain a featured book, and country listeners occasion-ally sent money with a request for the book in return, indicating a comfort-able trust between listeners and broadcaster (as well as the shortage of book sellers outside the capital cities).[50]

Scribe was generally praised, like Captain Stevens, for the skill with which he took on the voices of different characters.[51] For Scribe, character-ization was a key feature in determining the suitability of a book for reading on the radio. He wrote in 1938:

> Descriptive scenes should be given in the form of dialogue, characters should be well drawn; action is necessary and is much preferable to long descriptive scenes which only tend to delay the development of the plot.[52]

Scribe was known by the ABC to propose to authors that he make changes to their text to facilitate greater dialogue.[53] Clearly the versatility of his voice was adept at handling this, and it raises a tantalizing question about the manner in which Scribe voiced Indigenous characters. His readings of Upfield's books included the first three of the 'Bony' mysteries, the series in which Upfield developed the character of the part-Aboriginal, university educated detective, Napoleon Bonaparte.[54]

There is evidence of Scribe's presentation of women characters, which provoked a number of complaints, such as this from a woman listener in New South Wales:

> [T]o all his female characters he gives a cracked and quavering voice like that of an old lady of 80. I simply could not listen to it, and I do like to hear the stories.[55]

Internal ABC correspondence shows this was discussed at the senior level, with the Federal Controller of Programmes proffering the view that Scribe was 'an extremely bad female impersonator' and requesting that the mat-ter be discussed with him.[56] Women authors figured within Scribe's selec-tion; they included the writers Miles Franklin, Ernestine Hill, Hilda Bridges, Kathleen Jenner, Irene Rix Weaver and Myra Morris, amounting to about a quarter of the books he broadcast between 1931 and 1942. Available evidence of other storytellers' selections certainly includes books written by women.[57] However, throughout the 1930s and early 40s, there is no evidence of a serialized book read by a woman outside a children's programme.[58] The ABC in Victoria frequently engaged two women readers, Beatrice Touzeau and Ruth Conabere, for the broadcast of short stories, but during discussions in 1938 about the proposed national relay of the morning story, it was explicitly stated that they would not be included as national serial readers.[59]

This is a revealing discrepancy, a situation where women could write the words but not deliver them, and it testifies to the cultural significance of the physical, embodied voice: at this time in Australia, the public voice was predominantly male and the majority of radio announcers were men.[60] Women announcers were generally only heard on 'women's programmes.'[61] In the course of the 1930s, radio announcing developed an intimate mode, which, as Lesley Johnson has explained, came about both from press publicity about individual announcers and from their cultivation of on-air idiosyncrasies that had the effect of 'catching listeners through complicity, a position of inside information.'[62] The performance of dialogue by the radio storytellers played a similar game with the audience, drawing on the emerging conventions of radio listening where it is by the sound of a voice that the listener knows who is present. However, the voice that is central to this moment of play is also engaged in 'an exercise of power.'[63] To listen to a book reading, whether in a shared space with the reader or over the radio, is to accept the authority of the reading voice. It is to give up to—or perhaps to give in to—the demands it makes of time and attention. Radio storytellers at this time were presented as fully fleshed individuals rather than anonymous voices. The absence of women readers of book serializations over the course of the inter-war period reinforces the restrictions placed on women's ambitions to participate in public life on a footing equal to men and underscores the narrowness of the broadcasting conventions of the day.

Post-War Storytelling

The role of the Australian radio storyteller diminished over the war years. After funding cuts in 1942, the ABC decided to axe the morning serial, and many of the commercial slots were discontinued. In the years immediately after the war, public and commercial radio broadcast weekly short stories and daily drama serials, but the serialized reading of novels appeared a thing of the past despite the continuing publication of serials in many Australian newspapers into the 1950s.[64] When book readings did return—only on the ABC—it was as if a new chapter was beginning. Pre-war, storytellers could be found throughout the day in sessions aimed at general audiences and for the purpose of entertainment. The resurrection of the book serial in 1948 was in a designated women's programme with an 'improving' purpose. This was a vulnerable place in the schedules, but a later shift into gender-neutral territory ensured its survival. These changes reveal the impact of decisions around the framing of radio programmes, from which even a session as basic as a book reading was not immune.

The process began in January 1946 when Prime Minister Ben Chifley opened a four-day conference on radio and education convened by the ABC. The agenda was broad, with sessions on adult education and schools' broadcasting, as well as radio's engagement with culture and democracy. In one of the opening talks, Sydney University's Director of Tutorial Classes, Dr W.G.K. Duncan, discussed communication by print as opposed to that of radio, proposing the terms 'eye language' and 'ear language': he argued that communication based on listening needed to be taken more seriously, and suggested a revival of storytelling as a means of cultivating the ear.[65] The following day, the writer Vance Palmer spoke on 'Literature and Drama in Radio,' in which he distinguished talks *on* literary matters from the direct presentation *of* literature to the audience. Palmer was an advocate for the latter, arguing that the experience of listening to literature had the potential of lasting impact, as it 'may leave some enriching deposit in the mind or imagination, fertilising it with strange words, visions or points-of-view.' He went on to say 'a passage of literature has the power of scattering seeds that may germinate.'[66]

Palmer's words struck a chord with the ABC's National Talks Advisory Committee, a body composed largely of university academics. Minutes of their first meeting in 1946 record that 'several members expressed the view that a serial story would be an attractive item in the daytime pro-gramme.'[67] At the time, this committee had responsibility for the ABC's morning women's session, a subject of almost continual discussion.[68] Women's sessions had been established across commercial and public radio during the 1930s and were generally a mix of music and talk related to traditional 'feminine interests': 2UE launched 'Recipes and Rhythm' in 1936 and 2GB's flagship morning programme was 'Banish Drudgery.'[69] However, from 1942 the commercial radio mid-morning programmes began to include drama serials.[70] These were targeted overwhelmingly at a female audience, with listener allegiance bolstered by well-placed print promotions and opportunities to participate in the life of the dramas (for example, the offer of a dress pattern for the going-away frock of a leading character).[71] Audience research showed the ABC's mix of talks and music was a poor fit with the female audience despite successive alterations.[72] In January 1948, it was decided to revamp the session along 'modern magazine lines' and to precede it with a serial book reading, billed as 'the women's serial.'[73] The first book was the novel *Wind on the Water* by Myra Morris, published in 1938;[74] the *ABC Weekly* gave it a splash of publicity with photographs of the author and a full-page promotion for the reading.[75]

The 'Women's magazine' was described as 'a new ABC session for women of discernment.'[76] This suggested a listening subject for the book reading who was not just female and interested in 'feminine concerns,' but also 'discerning'—as if the ABC had identified an audience looking for something different from the standard commercial fare. There was some justification for this; Lesley Johnson and Justine Lloyd have argued that there was a prevailing tension about women's place in the post-war world, springing from a perception that women should be 'connected to, but not in, the public world.'[77] A surfeit of domestic chores was acknowledged as a recipe for boredom, and radio was perceived as being able to lighten the load with appropriately stimulating material. With the 'women's serial,' the one-to-one intimacy of a single voice was pitted against the escapism of a fully cast soap opera; by inference, the book reading acquired an exclusive aura—and a particularly feminine one at that. Before the war, the ABC national book reading was one among many, but the post-war decline of storytellers left the ABC's broadcast in the spotlight. Tal Ordell, who had been one of the most popular commercial storytellers, died in the same year as the women's serial started. This left only one survivor among the better-known pre-war cohort, Si Meredith, and publicity for his programme shifted to the women's pages of newspapers.[78] It was as if listening to a book being read aloud was no longer an activity suitable for a man.

In its publicity for the launch of the women's serial, the ABC had made no mention that the initial reader was Bebe Scott, nor of the fact that it was the first time a woman would read aloud nationally a complete novel. Scott was an accomplished and well-known radio actor, and she read the first three serializations of 1948, all of which were written by women. However, the opportunity for women to be the principal writers *and* readers of books on air was short lived, despite the reading's continued association with the women's session. Across 1949, readers were increasingly male, with women amounting to two-thirds of the featured authors. Audience research left ABC management scratching their heads, as it was clear that the 'Women's magazine' had a lower audience than the book reading attached to it. Their solution was to move the reading to a separate timeslot and replace it with a morning drama serial.[79] The audience for the 'Women's magazine' continued to languish, but the book reading flourished.[80] Its new timeslot, before 9 o'clock, was certainly more convenient for anyone going out to work, but arguably it was also *culturally* more comfortable for any men who were inclined to listen.

From 1952, a number of newspaper articles confirmed men among the book reading audience: in September 1952, the Brisbane *Sunday Mail* recommended the morning story of the moment, whose reader—Queenslander Ray Barrett—was no doubt known to the paper and its readers:

> We are not so blasé, so adult, that we've outgrown our pleasure in hearing a good story-teller, a man with an understanding and a voice to make a book live.[81]

The article notes that storytelling is limited on radio, and suggests replacing 'indifferent music programmes by the reading of a bestseller that would not otherwise be enjoyed by the majority.'[82] In November that year, the ABC's State Programme Director for Queensland proposed moving the morning reading to 6.45 pm, as 'the more mixed evening audience would presumably include many males who might welcome the opportunity of hearing good readings of popular novels.'[83] In 1954, the Adelaide *Mail* printed a plea by its radio correspondent John Quinn for book readings at a more accessible time.[84] There had been no expression of views related to male listeners while the book reading was branded as part of the women's session.

From the early 1950s, women accounted for about a quarter of the reading's writers and readers, until the balance began to move towards equal numbers in the 1970s. The readers were accomplished local actors, occasionally well known though often less so: post-war there was no regular cast. The commitment to Australian writing had been an early feature of the reading, and policy statements in the 1950s confirmed the rationale of the programme was to provide a broad range of listeners with the most recent Australian writing.[85] The 'Morning serial' became one of the ABC's longest running programmes, broadcast on local and regional stations at 8.45 or 8.50 am until the 1990s.[86] An evening transmission, simply called the 'Book reading,' was broadcast nationally until it was axed for financial reasons in 2013; it concluded with the short story 'A Happy Ending' by Rosa Praed, read by Tracy Mann.[87]

Conclusion

Book readings are small programmes, but their development since the 1920s demonstrates that they have a speaking part in the history of radio in Australia. The large number of book readings on early radio indicated a keen desire by listeners for narrative, both for stories that are not necessarily those of everyday life and whose duration over many weeks allows an immersive listening experience. Later book readings confirm the enduring nature of this appetite. There is a case for further scrutiny of the particular books that were presented to listeners in the past and, where possible, the responses listeners had to them, as well as a broader transnational analysis of radio book reading elsewhere. However, the focus of this investigation is the difference between book readings on Australian radio before and after the war, and that difference rings out with the sound of the reading voice. It is worth noting that there is something almost brazen about the role of reading in these broadcasts. The greater part of spoken word radio has involved the reading of a prepared script, but almost from the beginning speakers have been advised to conceal this fact.[88] They are urged to write in a manner specially suited to radio and to annotate scripts with the aim of convincing listeners that they are *speaking* to them. In the case of a book reading, the reader makes no such claim. The listener hears the voice of one person and the words of another: there is no pretence that it is anything but a performance. Furthermore, while radio over the last 40 years has set out to present the world in its sonic richness, radio book readings make no such effort. Book readings over this period have perfected a timeless, placeless performance with all extraneous sound removed—no page turns, no trace of lips, mouth or nose, barely a breath and any slip or hesitation surgically excised. As a remediation of a book, it has actively striven towards a dematerialized form.

It is a far cry from the 1930s when radio storytellers were reading 'live,' as opposed to recording, books of their own choosing. They were men, consolidating the gender conventions and power relations of the day; by the time women were admitted to their company, radio readers in Australia were on their way to becoming comparatively anonymous. The original radio storytellers were well known to listeners from their past broadcasts, their community activity and from press promotion which was occasioned by their very status as radio storytellers. They were extolled for their voice, its versatility and adaptiveness, with the boomerang effect of valorizing the practice of listening. A feature of the human voice is its uniqueness, and for philosopher Hannah Arendt it is a feature of the human condition, 'the unique shape of the body and sound of the voice.'[89] Arendt argues that knowing a voice is to know it in its uniqueness and that this knowledge in turn affirms the hearer's agency and particularity. This underscores the active participation a listener is engaged in when listening to a regular spoken word broadcaster. The opportunity for listeners to know the reader's voice was a characteristic of 1930s radio storytelling: the recognition of the voice stitches the listener's past into it, so that it is no longer clear whether the listener is engaged by the familiarity of the voice or the delivery of a particular reader. These are porous categories in themselves, as knowing a voice can influence any assessment of its capability. The storyteller offered listeners an aural experience more chameleon-like than that of the straight talk-giver; book reading created a flickering cast of characters that reinforced, and was reinforced by, the listener's knowledge of its source.

In the inter-war years radio was a new technology, a portal into a larger, faster, changing world, and the unreliability of its early reception made listeners aware of their tenuous grasp on the moment it offered—coupled with the understanding that if they missed an episode, the serial would go on without them. The reading voice through its sound offered a measure of stability, while its words were determined by the earlier technology of the printed page. The repetition of the readers' voices formed new rituals in listeners' lives, and the familiarity each listener had with the voice of a storyteller built on his or her individual memory of listening; personal experience was created alongside an awareness of mass communication. Listener response to book readings in pre-war Australia was loud and forceful; the cancellation of the programme in the twenty-first century raised barely a murmur, as if audiobooks can fill the gap. However, the commitment to the collective experience of the sound of the voice that arose from early radio book readings, a voice known inside and out, confirms a role for sound in the construction of individual agency and shared community, underscoring the complex negotiations radio gave rise to with the conditions of modernity.

Notes

1 Plato describes scenes of reading aloud in several dialogues (e.g., *Theatetus*, *Phaedo* and *Parmenides*), as well as refers to instances of silent reading by Socrates. While Alberto Manguel and others have declared that reading aloud was standard in the ancient world, this has been disputed—and arguably disproved—by Gavrilov 1997.

2 Hooper 1996, p. 17; Lyons 2001, p. 336; Damousi 2010, p. 142.

3 Hooper 1996; Lyons 2001, p. 341.

4 Edison 1878, p. 34.

5 Rubery 2011.

6 Radio book readings continue in the UK, US and NZ; this of course is in addition to radio services specifically for the blind. In Australia, the ABC axed its daily book reading in 2013, see note 87.

7 For example Avery 2006; Cohen 2009; Trotter 2013.

8 Hendy 2007 covers the editorial selections of the BBC's 'Book at Bedtime' and its later-night partner, commonly known as the 'Bonk at Bedtime.'

9 Australia has a dual broadcasting system, consisting of a large commercial sector and since 1932 the publicly funded Australian Broadcasting Commission (later Corporation); in the 1930s, radio listeners in Sydney could choose among six commercial stations and two run by the ABC. Country listeners might have access to two ABC stations and one commercial, while better reception at night often put a greater range of stations within reach. In addition, short wave frequencies for overseas services were routinely listed in broadcast magazines.

10 American publisher George Doran, quoted in newspapers across Australia, including *The Journal* (Adelaide), 31 March 1923, p. 1, *Morning Bulletin* (Queensland), 5 April 1923, p. 6, *Northern Star* (Lismore, NSW), 21 April 1923, p. 11. He went on to raise the need for a system for copyright fees.

11 Thompson 1966, p. 89; Griffen-Foley 2009, pp. 223–24. Book readings also figured in listener competitions (e.g., the challenge to identify a book from the reading of an extract: *Northern Star* [Lismore], 23 July 1927, p. 16; Thompson 1966, pp. 89–90, Griffen-Foley 2009, p. 224).

12 Johnson and Lloyd 1988, pp. 22–24; Damousi 2010, p. 252.

13 *The Listener In*, 1(1), 10 January 1925, front cover; the Empire Baby Competition, offering prize money of 200 pounds, was judged by a London-based panel of doctors and artists from a photograph and questionnaire, *Evening News* (Sydney), 13 March 1924, p. 8.

14 *Daily News* (WA), 17 September 1924, p. 7.

15 *Daily Standard* (Brisbane), 19 October 1926, 5.

16 Op. cit., 77.

17 Benson 1990, p. 75; see also Joy Damousi's account of the woman storyteller 'Mary, Mary' on Melbourne Radio 3LO (2010, p. 252).

18 *Sydney Mail*, 29 August 1928, p. 53.

19 Ibid.

20 *Daily Examiner* (Grafton), 8 May 1929, p. 8, *Wireless Weekly*, 28 June 1935, pp. 26, 18 January 1935, p. 24.

21 *Sydney Morning Herald*, 21 June 1934, p. 10.

22 For example, Captain Peters and Tal Ordell read published books, manuscripts and their own scripts.
23 Adelaide 2001, p. 92.
24 *Wireless Weekly*, 18 January 1935, pp. 26–30.
25 *Sydney Morning Herald*, 26 July 1934, p. 4.
26 *Sunday Mail* (Brisbane), 29 January 1939, p. 19; the *Maryborough Chronicle* in northern Queensland on 30 September 1941 describes Dick Turner as 'the 4MB storyteller,' reading a novel by Frederick Thwaites.
27 *Wireless Weekly*, 23 February 1934, p. 15.
28 Si Meredith was written up in 1940 as having read 100 novels in front of the microphone from 1933, *The Cumberland Argus and Fruitgrowers Advocate* (NSW), 3 April 1940, p. 8; *Wireless Weekly*, 21 June 1935, p. 17, photo of Ronald Morse.
29 'The renaissance of storytelling', *The Bulletin*, 15 August 1934.
30 *Wireless Weekly*, 18 January 1935, p. 22, listed as 'storyteller', 2UE. Stevens died in 1937, having been honoured in August 1935 with a testimonial programme at the Sydney Town Hall put on by the combined commercial radio stations in Sydney, *Sydney Morning Herald*, 24 August 1935, p. 8.
31 For example *Sand Dunes* by Warwick Deeping, read by Captain Stevens, *Wireless Weekly*, 26 April 1935, p. 35; other writers of stories in this run included A. Quiller-Couch, Stacy Amonier, E. Temple Thurston and Harold Wimburg.
31 *Wireless Weekly*, 5 April 1935, p. 71.
32 Ibid.
33 *The World's News* (Sydney), 21 February 1934, p. 18.
34 Ibid.
35 'While he swishes peas round the bottom of a cardboard box in front of "the mike", you get the illusion of murmuring waves on the sea shore,' ibid.
36 Ibid.; this applied similarly to Tal Ordell, Si Meredith and 'Scribe.'
37 Benjamin 1996, p. 148.
38 Barthes 1977.
39 Ellis Price, *Wireless Weekly*; and Si Meredith had been 'Uncle Si'; Tal Ordell *Radio Pictorial of Australia*, 1 October 1935, p. 18; Lesley Johnson notes that some figures associated with children's programmes had a following that included adults as well as children (Johnson 1988, p. 46).
40 *ABC Weekly*, 2 December 1939, p. 50.
41 This is thoroughly presented by Lindsey 2005.
42 *Listener In*, 28 July 1934, p. 9.
43 *ABC Weekly*, 2 December 1939, p. 50; he was also an Australian World War I digger, whose commentaries for Anzac Day marches were nationally broadcast.
44 NAA B2111/1, 5 May 1938.
45 op. cit., 14 April 1938.
46 Lindsey 2005, p. 138.
47 NAA B2111/1, 13 January 1939; Lindsey 2005, p. 130; Scribe averaged one Upfield book a year from 1935 to 1942, with Upfield receiving an initial fee of 10 guineas.
48 Lindsey 2005, pp. 94, 127, 132.
49 ibid., pp. 131, 134.
50 Listeners' mail summaries, ibid.
51 *The Age*, 2 August 1938, p. 10.
52 Letter from Williams to Proud, Manager ABC Victoria, 19 December 1938, NAA B2111/1 TKS 10; this followed a suggestion in 1938 by Tasmanian writer Roy Bridges to write a serial expressly for broadcast. The ABC declined due to cost. Later in 1951, a serial writing completion on an historical theme was organized as part of the celebrations for the Jubilee of Federation.

53 NAA B2111/1 TKS 10, national relay of morning serial.

54 In his autobiography, Upfield describes meeting the mixed-race man sometimes thought to be the model for Bony; the first encounter is recorded as occurring through the man's voice, 'soft and liquid and finely modulated'; Lindsey 2005, p. 85.

55 NAA B2111/1/TKS 10, morning serial story—national relay, 7 June 1939.

56 Ibid., memo from Keith Barry, 24 May 1939.

57 NAA, op. cit., 21 July 1942; *The Advertiser*, 13 January 1944, p. 8.

58 'Elizabeth' was included as one of 2BL's 'official fictioneers' in 1934 by *Wireless Weekly*, but her name is not included in a report of the same story by *Australian Radio News*, 23 February 1934, p. 9, nor does she figure among the profiles of storytellers such as Tal Ordell, Ellis Price, Si Meredith, Scribe and Captain Stevens.

59 NAA, op. cit.

60 Damousi 2010, p. 251.

61 This does not include the presentation of talks and participation in discussions, as women participated in these in the same time slots as men, albeit less frequently.

62 Johnson 1988, p. 95.

63 Connor 2000, p. 23.

64 Short stories were on the ABC national network (NAA MT395/1: 'Tuesday short story') and 2CH (*ABC Weekly*, 15 May 1948, p. 19).

65 ABC 1947. Radio in Education Conference 1946, pp. 20–31.

66 Palmer 1946, p. 68; the word 'germinate' is also used by Walter Benjamin in his essay 'The Storyteller' (1996, p. 148). Benjamin writes of the nature of the power of a story told by a teller, contrasting it with information: 'information does not survive the moment in which it was new,' whereas a story 'preserves and concentrates its energy and is capable of releasing it even after a long time . . . it is like those seeds of grain that have lain for centuries in the airtight chambers of the pyramids and retained their germinative powers to this day.'

67 NAA SP1474/1, National Talks Advisory Committee, Minutes, January 1946.

68 Inglis 1983, p. 169; see also Johnson and Lloyd 2004, Chapter 5, particularly pp. 127–41.

69 *The World's News* (Sydney), 2 September 1936, p. 30.

70 *Big sister*, broadcast from 2 February 1942, 2UW; Lane 2000, pp. 63–64.

71 *The Advertiser* (South Australia), 25 November 1944, p. 5; the 'imagined communities' that listening to serials gave rise to among their audiences have been discussed by Megan Blair.

72 NAA SP1474/1, National Talks Advisory Committee, Minutes, Box 2, 19 January 1946.

73 NAA, op. cit., 27–29 January 1948; and NAA SP724/1, National Women's Session, Box 36, 63, 20 January 1948; *ABC Weekly*, 3 April 1948, p. 18.

74 *Wind on the Water* had been a *Women's Weekly* recommendation in 1938; it had been read nationally by Scribe in 1940.

75 *ABC Weekly*, 3 April 1948, pp. 18, 47.

76 Ibid., 47.

77 Johnson and Lloyd 2004, p. 139.

78 *Sydney Morning Herald*, 11 September 1952, p. 7, and 23 April 1953, p. 7.

79 NAA SP1474/1, National Talks Advisory Committee, Minutes, Box 2.

80 NAA SP341/1/2.3, Part 1, Serials, Box 10, Memo Barry, 21 May 1952.

81 *Sunday Mail* (Brisbane), 7 September 1952, p. 15.

82 Ibid.

83 NAA SP411/1/2.3, Part 2, Serials, Box 7, Queensland Programme Director's report, November 1952.

84 *The Mail* (Adelaide), 22 August 1954, p. 35.
85 NAA, op. cit., 13 July 1954 and 29 Oct 1957; overseas writers were featured from the 1960s, and the twenty-fifth anniversary of the book reading in 1973 was publicized with a list of authors none of whom was Australian; later years included many books in translation, but the majority of those read were by Australian writers.
86 Other book reading series, for instance series of autobiographies, came and went during the years 1958–2011.
87 The last edition of 'The Book Reading' on ABC Radio National was broadcast on 28 January 2013, produced by Anne Wynter.
88 Carney 1999, p. 32.
89 Arendt 1958, p. 179.

References

ABC 1947. *Radio* in Education Conference 1946.

Adelaide, D., 2001. Chapter 5 'How did authors make a living?', in M. Lyons and J. Arnold, eds, *A history of the book in Australia 1891–1945: A national culture in a colonised market*, vol. 2, Brisbane: University of Queensland Press, pp. 83–103.

Arendt, H., 1958. *The human condition*, University of Chicago.

Avery, T., 2006. *Radio modernism: Literature, ethics, and the BBC, 1922–1938*, Aldershot: Ashgate.

Barthes, R., 1977. 'The grain of the voice', in *Image, music, text*, London: Fontana.

Benjamin, W., 1996. 'The storyteller', in *Selected writings*, vol. 3, Cambridge, MA: Belknap Press.

Benson, R. J., 1990. 'The establishment and early development of broadcasting in Queensland', BA Honours dissertation, Griffith University.

Carney, M., 1999. *Stoker: The life of Hilda Matheson*, Llangynog: privately published.

Cohen, D. -R., 2009. Chapter 8, 'Annexing the oracular voice: Form, ideology and the BBC', in D. -R. Cohen, M. Coyle and J. Lewty, eds, *Broadcasting modernism*, Gainesville: University Press of Florida, pp. 142–57.

Connor, S., 2000. *Dumbstruck: A cultural history of ventriloquism*, Oxford: Oxford University Press.

Damousi, J., 2010. *Colonial voices: A cultural history of English in Australia, 1840–1940*, New York: Cambridge University Press.

Edison, T., 1878. 'The phonograph and its future', *North American Review*, 126, pp. 530–36, reprinted in T. Taylor, ed., *Music, sound and technology in America*, Durham, NC: Duke University Press, pp. 29–36.

Gavrilov, A. K., 1997. 'Techniques of reading in classical antiquity', *The Classical Quarterly*, 47(1), May, pp. 56–73.

Griffen-Foley, B., 2009. *Changing stations*, Sydney: UNSW Press.

Hendy, D., 2007. *Life on air: A history of radio four*, Oxford: Oxford University Press.

Hooper, W., 1996. *C.S. Lewis: A companion and guide*, San Francisco: Harper.

Inglis, K. S., 1983. *This is the ABC: The Australian broadcasting commission, 1932–1983*, Melbourne: Melbourne University Press.

Johnson, L., 1988. *The unseen voice*, Oxford: Routledge.

Johnson, L. and Lloyd, J., 2004. *Sentenced to everyday life: Feminism and the housewife*, Oxford: Berg.

Lane, R., 2000. *The golden years of Australian radio drama*, vol. 2, Australia: ScreenSound Australia.

Lindsey, T. B., 2005. 'Arthur William Upfield: A biography', PhD thesis, Murdoch University.

Lyons, M., 2001. in M. Lyons, ed., *History of the book in Australia*, vol. 2, St Lucia, Qld: University of Queensland Press.

Palmer, V., 1946. Australian Broadcasting Commission. *Radio* in Education Conference (January 20–24, 1946: Canberra) n.p.

Rubery, M., ed., 2011. *Audiobooks, literature and sound studies*, Oxford: Routledge.

Thompson, J., 1966. 'Broadcasting and Australian literature', in C. Semmler and D. Whitelock, eds, *Literary Australia*, Melbourne: F.W. Cheshire, pp. 89–117.

Trotter, D., 2013. *Literature in the first media age: Britain between the wars*, Cambridge, MA: Harvard University Press.

I would like to thank Paula Hamilton for her insightful and generous assistance in preparing this chapter for publication.

4 Lindbergh's Voice

David Goodman

Aviator Charles Lindbergh was the most popular and charismatic orator in the 1939–1941 debate about United States entry into World War II. He became the public face of isolationism, the movement arguing that American interests were best served by staying out of what Lindbergh evocatively called 'these eternal wars in Europe.'[1] The debate about the war was divisive and often bitter, but featured high levels of citizen engagement. While the isolationists were at first in the majority, their support slowly diminished as news from Europe and China shifted more Americans into the interventionist column. Although there were Americans of all classes on both sides of the debate, isolationists prided themselves on being plain people, without—they often claimed—the wealth and privilege of their opponents. It was quite fitting therefore that the leading isolationist orator should be above all else a plain speaker.

Lindbergh began his public speaking on the war with radio addresses, and then moved to speaking in front of live and large audiences. President Roosevelt's scriptwriter, the playwright Robert Sherwood, recalled Lindbergh as 'an extremely eloquent crusader for the cause of isolationism,' and as 'undoubtedly Roosevelt's most formidable competitor on the radio.'[2] *Life* magazine provided testimony in August 1941 about the response to his public speaking:

> Another man who recently spoke from the same platform with Lindbergh and who has had considerable political experience said: 'Men are symbols, whether they want to be or not. At Madison Square Garden the applause for Lindbergh and for Wheeler was about the same in volume—but in quality it was entirely different. Lindbergh evokes a fervor, a tension . . . Hitler has the same thing; Roosevelt has it sometimes; Huey Long used to get it, and Coughlin, occasionally. I know Lindbergh doesn't seek it especially, and does nothing to stir it, but it is there.'[3]

Lindbergh's interventionist opponents lamented to themselves their lack of a headline speaker with this kind of crowd-pulling and persuasive capacity.

President Roosevelt, perhaps the only speaker who could have equalled Lindbergh's platform power, remained largely above the fray.

Lindbergh's speeches of course also divided Americans; amidst increasingly frequent charges that he was an appeaser at best and a Nazi sympathizer at worst, his speeches provoked intense hostility as well as admiration. I will concentrate here, however, on the positive appeal of his speeches, because that seems to pose the greater puzzle for a sound-conscious history.

It is difficult now to *hear* what was so compelling to so many Americans about Lindbergh's speeches. Listen today to the few recordings of the major Lindbergh speeches: they sound flat, unexciting, almost pedantic. Run some of the recordings through a tone-analysing programme and it is clear how repetitive Lindbergh's intonation was—the same falling cadence at the end of each sentence, regardless of sense, regardless of whether the sentence was a question or a statement.[4] Even when he had graduated to speaking in front of packed and adoring mass rallies, Lindbergh was closer to a careful reader of his laboriously written scripts than an impassioned, fluent orator. He lacked the performative instincts of a Roosevelt, whose artful pauses even conveyed drama and meaning.[5] In a nation with traditions of powerful oratory, it is not easy to hear or understand now from the surviving recordings why Lindbergh's stilted, formal, cautious radio speeches qualified him then as a great speaker. Novelist Philip Roth articulates this contemporary response in his 2004 novel *The Plot Against America*, describing Lindbergh's voice as 'undistinguished' and his speech as 'unadorned and to the point, delivered in a high-pitched, flat, midwestern, decidedly un-Rooseveltian American voice.'[6]

So what are we missing? Why can we not hear *in his voice* the reasons for the excitement Lindbergh generated? Back in 1941, even a listener who deeply disagreed with Lindbergh wrote to say: 'Your voice is clear and beautiful. Your diction is perfect. You have a great future in America.'[7] The historical question has to be—what were they hearing that we are in danger of not hearing? How do we reconcile the memory of Lindbergh's great oratory with the sound we can still hear today in recordings?

It might of course be suggested that the effectiveness of Lindbergh's speeches had more to do with the man than the voice. Lindbergh was a man who—because of his 1927 solo flight across the Atlantic and the 1932 kidnapping of his baby—had become a celebrity of almost unprecedented standing in the early mass media age, the object of what psychoanalyst A. A. Brill labelled the 'wildest display of hero worship ever accorded a single human being.'[8] Some listeners to his anti-war speeches said they could hear the 1927 air hero in his voice and sentiments: a New Jersey listener wrote that, after listening to Lindbergh's 'beloved voice' over the air, she imagined 'again it was the Lone Eagle who dared the unknown terrors of a lonely flight across the ocean wastes, now flinging his challenge to the frowning giants who threaten the destruction of our Country!'[9] More commonly, pro-Lindbergh Americans wrote to praise the content of his speeches, what one listener described as his 'brave words.'[10] The excitement that Lindbergh's speeches generated had to do with who he was, but it was more fundamentally about what he said and how he said it.

It might be suggested that if then it was the ideas that mattered, Lindbergh's persuasiveness could have worked as well in written as in aural mode. The speeches were reprinted in multiple places, and many Americans presumably read rather than heard them. But it was radio that was most responsible for Lindbergh's following, and the speeches were sonic more than literary events. The largest national isolationist group, the America First Committee (AFC), understood this. Concerned to get Americans to listen to rather than read the speeches, it organized prior newspaper publicity and hosted events to draw attention to them. An AFC organizer told Lindbergh in February 1941 that he planned 'a nationwide chain of radio "parties" both in private homes and in public meeting places to increase the size of the audience.'[11] *Listening together* to the broadcasts of live Lindbergh speeches seemed to AFC organizers the optimal way to generate enthusiasm and support.

We can apprehend the importance of the radio broadcasts as sound events and experiences in the way that members of Lindbergh's radio audience reported back to him in some detail about their listening. 'You came through just swell,' wrote a St Louis woman, 'and we understood every word perfectly.'[12] Sympathetic listeners wrote about how Lindbergh's words affected them, about the excitement and pleasure they experienced before their radios. A Wisconsin woman explained: 'I can only hear you speak as your voice carries over the radio. And it does my heart and soul good to listen to your broadcast.'[13] A woman from Cincinnati enthused that 'words cannot express how good we feel when we hear your voice over the radio. You speak with such conviction and understanding.'[14] A woman from San Francisco reported coming close to imagining herself in the live audience: 'I was so thrilled that it seemed almost as though I had been transported there to Cleveland, Ohio and could scarcely refrain from applauding the radio. I think it is by far the greatest speech I ever heard.'[15] It is no coincidence that so many of the warmest responses came from women. Lindbergh's particular appeal to women had been noted since his heroic 1927 flight, after which psychoanalyst A. A. Brill explained, somewhat coyly, in *Popular Science* that Lindbergh's unmarried state allowed women to 'think of themselves as the possible Mrs. Lindbergh.'[16] By 1940 it was arguably not Lindbergh's sex appeal, but his rationality and control that were most appealing, to men as well as to women.

From Studios to Mass Rallies

Lindbergh was by 1939 a national public and political figure who would rapidly become an effective radio broadcaster. His growing political capital rested to a considerable extent on the effects—emotional, persuasive, communicative—of his broadcasts. The message of his speeches remained fairly constant—a plea for American independence; a reassurance that the oceans still provided a buffer for American defence; a persistent request that Americans decide on the war question with reason rather than emotion; hints that foreign or un-American people or forces were behind the push towards war; a confidence that the real American people were solidly behind him and that if American democracy survived, the United States would not join the European war. However, Lindbergh's speeches changed in form and context because of his growing success and reputation as a speaker, and his consequent move from studio-based to live, on-location broadcasts from mass rallies. We need some sense of the cycle of speeches he delivered and of how the context changed between September 1939 and November 1941. I discuss here just a few of the more significant speeches.

Lindbergh's first speech on the war, on 15 September 1939, was a 15-minute address at 10.45 pm EDT on a Friday evening, carried on all three major networks, thus 'creating an audience of a size usually only enjoyed by the President.'[17] He broadcast from his hotel room, 'speaking into a battery of microphones.'[18] The speech received a relatively bipartisan and generally positive response. Biographer Scott Berg reports that Lindbergh was unhappy with his 'high pitched and flat delivery,' but comments astutely: 'strangely, his unimpassioned tone accentuated his sincerity.'[19] On 19 May 1940, Lindbergh made another studio broadcast on 'the air defense of America'; the surviving recording reveals that he spoke slowly and precisely—about 100 words a minute, slower even than Roosevelt's average fireside chat speed.[20]

On 4 August 1940, Lindbergh made his first live broadcast anti-war speech at Soldier Field in Chicago. The stadium was only half full, but 40,000 people was a large crowd, and the stadium offered a very different acoustic environment to the radio studio. Lindbergh noted that people were constantly coming up to ask him questions, making it impossible for him to follow what was being said. His wife Anne Morrow Lindbergh helped write his speech, which contained some optimistic thoughts about the future of civilization: 'Let us offer Europe a plan for the progress and protection of the western civilization of which they and we each form a part.' Anne, listening at home, wrote that she was 'moved' when Charles said: 'I prefer to say what I believe, or not to speak at all. I would far rather have your respect for the sincerity of what I say than attempt to win your applause by confining my discussion to popular concepts. Therefore, I speak to you today as I would speak to close friends rather than as one is supposed to address a large audience.'[21] Coming so early in the cycle of his anti-war speeches, that note of sincerity and profession of lack of artifice helped set the frame within which Lindbergh's speeches would be heard and appreciated. This was Lindbergh's first experience of speaking about the war to a large live audience. He found the crowd's propensity to applaud 'at every opportunity' to be 'rather disconcerting when one is not used to it.' But overall he liked the experience, noting that he found it 'much easier to speak to an audience than to microphones alone.'[22]

Criticism of Lindbergh's position intensified from his Monday 14 October 1940 studio broadcast (also filmed for the newsreels), 'A Plea for American Independence,' his fourth radio talk on defence issues. In this talk he focused on leadership in the context of the elections: 'The fact is today that we are divided; we have not confidence in our leaders. . . . They harangue us about democracy, yet they leave us with less knowledge of the direction in which we are headed than if we were citizens of a totalitarian state.'[23] Vice-presidential candidate Henry A. Wallace responded to the speech by labelling Lindbergh the nation's number one appeaser.[24] Soon after, on 30 October 1940, Lindbergh spoke again in front of a live audience, at Woolsey Hall at Yale, giving his longest address yet. He expected heckling, but received none: 'They listened attentively during the entire half hour, and it seemed everyone in the hall was clapping after I finished!'[25]

In 1941, Lindbergh addressed a series of mass rallies as headline speaker. These rallies created a quite different sonic environment, one increasingly familiar to radio listeners worldwide. Hitler and Mussolini broadcast only from live gatherings so that the sound of the crowd became part of the broadcast event for listeners at home.[26] Birdsall notes that this 'adoption of a public mode of address appealed to Germans simultaneously as a collective of listeners rather than as the anonymous, individual subjects they, in fact, usually were.'[27] In the United States too, broadcast crowd sounds made a big impression on listeners at home. Letters evidence that audiences were acutely aware of listening-in to a live event and that they frequently commented on how the crowd sounded and how they interpreted its response.

On 23 April 1941, Lindbergh spoke for America First at the Manhattan Center, saying that the United States was being led to war by a minority of its people. There were thousands of people outside as well as inside the auditorium.[28] Anne Lindbergh wrote of this speech: 'The crowd galvanized by him, silenced, turned to him. And when he started to read, slowly, with emphasis, I felt his great power and strength and I watched that crowd looking at him, with faith, with undivided attention, with trust.'[29] On 23 May 1941 there was an America First rally at New York's Madison Square Garden—there were 22,000 inside and again many thousands outside unable to get in. Listening around the nation on the Mutual network, you would have heard the roar of the crowd as Lindbergh was introduced by John Flynn (chair of the New York AFC chapter) as 'Citizen Charles A. Lindbergh.' There was prolonged cheering and applause, and the crowd began slowly to chant: 'We want Lindbergh. We want Lindbergh.' The crowd noise and response led him to pause between sentences, slowing his delivery to not much more than

100 words per minute; the intonation was flatter, with less fall in pitch at the end of sentences than in his studio broadcasts.

On 11 September 1941, Lindbergh made the speech that provoked the greatest response—positive and negative—of all his anti-war speeches. In Des Moines, Iowa, he named the three groups he said were agitating to get the United States into the war: the Roosevelt administration, the British and the Jews. That speech culminated the cycle of Lindbergh's war speeches— making explicit what in earlier speeches had been merely implied—and bitterly divided Americans as it did so. While there were isolationist Jewish Americans, Lindbergh was no doubt correct that most favoured intervention. His speech, however, provoked condemnation from almost all sides, because it so starkly threatened to license turning a debate about the national interest into a divisive, identity-based scrutiny of loyalties. His naming of Jewish Americans as war agitators crossed a line—delighting some of his more passionate followers, but alienating most of what remained of his liberal nationalist support by appearing to endorse the typical anti-Semitic exaggeration of the power, wealth and influence of the Jewish population of the nation.[30]

'Speaking Is Not My Vocation'

Lindbergh in these war speeches sounded cautious and precise, not like a naturally confident orator. The nominees of the two major parties for president in 1940 were both markedly more fluent public speakers. Franklin Roosevelt was widely regarded as a natural orator, although he relied more than any previous president on speechwriters.[31] Republican Wendell Willkie was also perceived as a natural speaker; many times he drew attention to this capacity by throwing away his notes while speaking.[32] These were the men against whom Lindbergh pitted himself in oratorical combat, usually without naming them. Instead, he commented on leaders and leadership generally, calling for example in his 13 October 1940 radio address for 'leadership which is entirely and unequivocally American.'[33]

In contrast to these natural orators, Lindbergh understood himself as a speaker who needed a lot of time for preparation. That he wrote his own speeches, despite some claims at the time to the contrary, is obvious from the several drafts of each preserved in his papers. Beginning with notes, he progressed to handwritten drafts, and then to corrected and revised typescripts; two live-in secretaries were available to help with such tasks. Lindbergh felt he needed this time to prepare. He declined an invitation to speak in September 1940, explaining that 'I would not want to address one of the meetings you suggest without preparing my address very carefully— I have not the gift of being able to speak well extemporaneously.'[34] The next month he declined an invitation to speak on George Denny's high profile America's Town Meeting of the Air on NBC 'because I have had no experience in taking part in impromptu discussions of this type and feel that this would be a particularly bad time to begin.'[35] In April 1941, after several months of public speaking, he confided to Chicago's Colonel McCormick that 'the only way I can make up for my inexperience as a speaker is by devoting what seems an inexcusable amount of time working on my addresses.'[36] Knowledge of Lindbergh's laborious speechwriting habits leaked to the press: 'He composes them, it is said, like so many high school graduation essays, and it takes all the tuck out of him. He is deadly in earnest—puts everything in his speech, and then expects the world to take it or leave it.'[37] To his sympathizers, however, that careful crafting paid off and his simple delivery evoked sincerity and lack of artifice. African American Republican Roscoe Conkling Simmons found a passage in Lindbergh's April 1941 Chicago speech to be 'eloquent.' 'Three things make eloquence,' he explained in his *Chicago Defender* column: 'feeling, speech, delivery.'[38]

It was not only in private correspondence that Lindbergh professed his un-suitedness for public oratory. This became a theme in the speeches themselves, a classical republican trope about the practical man who turned to political contestation only when his country was in danger. In St Louis in May 1941, Lindbergh told the crowd:

> I would far rather be flying in my country's Air Force than appearing on lecture platforms and making radio addresses. Those of us who are arguing against the war have nothing to gain except the welfare of our country. We speak only from the depths of our conviction. Most of us desire nothing more than to return to our private lives and occupations. But we know that this nation is being led into a major disaster, and we would be poor Americans, indeed, if we stood quietly by without even raising our voices in opposition.[39]

A few months later in Fort Wayne, he spoke on the same theme:

> In making these addresses, I have no motive in mind other than the welfare of my country and my civilization. This is not a life that I enjoy. Speaking is not my vocation, and political life is not my ambition. For the past several years, I have given up my normal life and interests; first, to study the conditions in Europe which brought on this war, and, second, to oppose American intervention. I have done this because I believe my country is in mortal danger, and because I could not stand by and see her going to destruction without pitting everything I had against that trend.[40]

This self-characterization holds one major clue to the puzzle of the sound of Lindbergh's voice. Lindbergh stated his absence of obvious skill for political oratory, and he performed it; he made his lack of 'natural' oratorical skill into a republican virtue. His slow and deliberate speaking became a demonstration of his control of his emotions, even when talking of the emotive issues of war and peace.

Impersonal as a Surgeon with a Knife

Like many of his generation, Lindbergh viewed emotions as unwanted and avoidable intrusions into an ideally passion-free public sphere.[41] In his first speech on the war, he set the tone for his broadcasts to come when he said: 'We must not permit our sentiment, our pity, or our personal feelings of sympathy to obscure the issue, to affect our children's lives. We must be as impersonal as a surgeon with a knife.' The success of this self-characterization is demonstrated in the reporting of this speech. *Life* described it as 'clear, concise and calm.'[42] A newspaper columnist called it a 'true, sentiment-shorn' account of the state of affairs overseas.[43] Lindbergh understood and successfully projected himself as a rational man in an emotional age. He noted in his journal entry for 10 October 1940: 'I am using logic at an emotional and illogical time.'[44] His public persona was that of the technical expert, the engineer. His constant implication and assertion was that only overly emotional empathy with the suffering of people in other nations could lead Americans into participation in a war that would be against their own interests.

Lindbergh's opposition of rationality to emotion resonated broadly within the isolationist movement. The isolationists constantly claimed to be rational, free from the pro-British emotion and war hysteria they feared were engulfing the rest of the populace. Their story was that the anti-war side had preserved its reason, successfully fought off emotional appeals. Historian James Schneider notes that 'admonishments against "war hysteria" were a staple of commentary nation-wide during the early days of the war.'[45] Amidst such concerns, Lindbergh's plain, flat, earnest voice sounded to sympathizers both rational and reassuring. The very repetitiveness of his inflection and cadence became a kind of aural reassurance that he was not being emotionally transported by his own words—like a Hitler for example, whose stadium oratory style would have been familiar to American radio listeners as his speeches were at various times broadcast (with English commentary) over the American networks.

Lindbergh's technical persona and emotional coolness resonated with many Americans at a time when emotional control was increasingly valued; historian of emotions Peter Stearns indeed identifies a 'new aversion to emotional intensity' from the 1920s.[46] Lindbergh supporters heard the cool rationality in his voice. Indiana member of the U.S. House of Representatives, Louis Ludlow, put it succinctly: 'I listened to your radio address and I think that yours is the voice of sanity in a storm of hysteria.'[47] The sympathetic public responded in similar terms. 'It gives us all hope to hear your voice,' wrote a New York woman, 'a voice that sounds sincere, honest and logical.'[48] Lindbergh's relative lack of animation and oratorical emotion became for such listeners aural confirmation of his reason and responsibility. A lawyer from White Plains, New York, wrote 'your address of last evening in Oklahoma was the greatest American speech of this generation. . . . No other man in this country has given so civilized, so penetrating, and so calm a presentation of these matters as you gave last night.'[49] A professional violinist from St Louis echoed back perfectly Lindbergh's self-characterization as rational and above emotion. 'Your speeches are refreshingly free from the emotional dramatics which characterize the outpourings of our pseudo patriots,' he wrote. 'It seems to me that a problem to be solved, whether it be in mathematics, physics, aerodynamics or the proper working of a government, should be approached with cool-headed impartiality and objectivity. It is here that the interventionists betray the weakness of their position; their arguments range from mouthing hysterical hatred to more subtle expressions of heart torn sympathy for oppressed peoples.'[50] Listeners, then, heard Lindbergh's un-oratorical, pedantic but precise voice as genuine, and as rational rather than emotional, and they knew that that was what was needed in an age of propaganda. His plain, sentence-by-sentence voice clearly sounded to sympathizers as if it spoke the unadorned truth—honest and sincere, calm and reasoned. A woman in Lincoln, Nebraska mentioned his 'courage, sincerity and honesty.'[51] A woman from Chicago reported, 'Your speech broadcast over the radio Saturday Night . . . was comprehensive, logical and fearless.'[52]

In June 1940 the isolationist *Chicago Tribune* reported on medical experts who advised that listening to Lindbergh's speeches helped combat hysteria. Dr Irving S. Cutter, dean of the medical school of Northwestern University, and Dr George W. Hall, neurologist and psychiatrist at St Luke's hospital, warned that the war was having a deep emotional effect on the population. 'Many persons tell me they wake up feeling fine, then suddenly remember Europe and the war and become greatly depressed. What these persons should do is keep their minds active on other matters. They should be interested in other people, in books, and in the outdoors in order to keep their minds off the holocaust in Europe.' Hall stated that 'he felt pronouncements of the sort made recently by Col. Charles A. Lindbergh were of value in keeping down hysteria.'[53] In this account, Lindbergh's cool voice became actually therapeutic as well as conducive to right decision making.

Anne and Charles on Speech and Feeling

Lindbergh approached speaking as a technician. In his journal he mostly recorded just the facts—the size of the crowd, the number of minutes he spoke (on target or not). Of the heady Madison Square Garden speech, he simply noted: 'I spoke for twenty-five minutes.'[54] Even amidst the adulation, he did not greatly enjoy the performance of public speaking or the exposure to crowds that it entailed. He wrote in his journal in May 1941 that he hoped that after the war he would never have to make speeches again: 'I would not mind speaking so much if I could take plenty of time to prepare an address . . . and then give it from a platform, without any elaborate ceremony, and say good night.' He appeared to think of public speaking in this way as the delivery of a message. We might even see an analogy here to his earlier career as an airmail pilot, delivering bags of mail. But in oratory of course, unlike in the postal service, the mode of delivery, and the performance of delivery, are crucial.

Anne wrote in her diary for 13 September 1939 that she agreed with Charles on the war—'I believed in him and his stand (no matter how hard it was emotionally to turn one's back on Europe).'[55] Her subsequent interventions in and comments on his speeches stemmed from that dual response—sympathy and intellectual agreement, but also a protective desire that he should convey an understanding that the turning away from Europe came at an emotional cost to many Americans. Lindbergh's Madison Square Garden speech began: 'We are assembled here tonight because we believe in an independent destiny for America. Such a destiny does not mean that we will build a wall around our country and isolate ourselves from all contact with the rest of the world. But it does mean that the future of America will not be tied to these eternal wars in Europe. It means that American boys will not be sent across the ocean to die so that England or Germany or France or Spain may dominate the other nations.'[56] Anne noted in her diary: 'He speaks slowly and deliberately. It sounded so calm and unemotional.'[57]

In fact, for this and the other speeches, Anne Lindbergh had carefully marked up his reading script in pencil with performance directions for him—offering cues about speed, emphasis and emotion. 'End and beginning should be spoken slowly with great emphasis + feeling. Middle portions . . . faster.' Against the sentence which said 'we will fight anybody and everybody who attempts to interfere with our hemisphere, and that we will do so with all the resources of our nation,' she wrote 'emphasize this sentence slow + loud.' Where Charles had written bluntly of the possibility of French and English defeat ('I knew that England and France were not in a position to win and I did not want them to lose'), Anne tried to inject the word 'feeling' ('because of the feeling I had for those countries I did not want them to lose')—again unsuccessfully.[58]

You might conclude that Anne thought Charles needed careful direction about speaking, and that she thought he was not a natural orator. You would be right. For his 15 September 1939 speech, Anne contributed a revised ending that injected some hope and human feeling—the plea that 'the gift of civilization' be protected because it was 'more important than the sympathies, the friendships, the desires, of any single generation.'[59]

She may have made the same sort of contribution to the 1940 Chicago Soldier Field speech, which also contained some more optimistic thoughts about Western civilization. She rewrote some of his Philadelphia speech. She recorded in her diary her relief at hearing when his speeches succeeded on radio. On 10 May 1941 he spoke in Minneapolis and she wrote: 'Hear C. speak on the radio. It is very good. It all comes over. And they seem very enthusiastic—long and loud applause. I am relieved.'[60] She worried about his delivery and his blunt statements (most strenuously warning him about the 11 September Des Moines speech and the consequences it was likely to have).

The line between cool and cold was always contested. The constant risk of Charles Lindbergh's self-presentation was that Americans would find him cold rather than cool. No one understood that risk more acutely than Anne Lindbergh. Chicago writer Carl Sandburg said of Charles Lindbergh in June 1941 that he had 'ice in his veins.' Anne was hurt by this characterization, 'shocked beyond words.'[61] She worked behind the scenes to make Charles more aware of the performative aspects of public speech and to inject feeling into his delivery. Charles was, however, resistant to rehearsing expression and emotion. He wrote in his journal: 'Spent most of the afternoon listening to Anne practice her talk. She reads and speaks much better than I do. I should be the one who practices reading my addresses; but I hate to read something I write more than once. I seem to lose spirit and feeling in the second reading.'[62] He most of all wanted control, and thus preferred studio broadcasts to live, and he distrusted newsreels, which could be edited to place his comments in unflattering context. He just wanted Americans to hear his emotionally cool, rational speeches without interference.

Conclusion

Charles Lindbergh was of course on the losing side of the war debate. His apparent lack of vigour in condemning Nazi atrocities, his calls for a national heart that was closed to the suffering of non-Americans, have not endeared this phase of his career to many subsequent historians.[63] But Lindbergh in these years generally succeeded in setting the terms within which his own hand-made rhetoric and public speaking became inspiring to millions of his followers. In order to understand this history, we have to try to listen as his supporters did, in order to comprehend how it was that they heard a kind of national salvation in his flat, precise, unexcited voice.

Anne Lindbergh understood another important thing. Charles Lindbergh's voice, despite its apparent lack of emotion, stimulated strong emotion in his listeners. She observed after the Des Moines speech that, while Lindbergh's goal was to get Americans to view the war situation dispassionately, 'just the opposite is happening—more passion is being aroused.'[64] His unemotional voice somehow provoked love—love of Lindbergh himself as a great patriot, and of the nation he sought to (re)define and defend. Agnes Hovde from Lindbergh's home state of Minnesota wrote to him in August 1941: 'We urge you, our Leader and our Voice, for we have none other, to persist as far as your strength and time allow in carrying our crusade to the uttermost; and to speak to us again as you spoke last night.'[65]

In order to understand what people like Agnes Hovde were feeling, we need to begin with what they were hearing. The history of sound has much to contribute to a history of the great debate about United States entry into World War II. If we only read the speeches and the newspaper commentaries, we lose some of the crucial context of this radio-age debate. The sounds of particular national voices, of the background noise of the mass rallies, heard over the radio, made citizens such as Agnes listeners-in on history. I have tried here to explicate the set of expectations that she and other pro-Lindbergh Americans had of leaders—that they would indeed possess a Voice that could lead the people, interpellated as unemotional patriots, to a clear sense of their collective self-interest.[66]

Notes

1 *New York Times* 1941.
2 Sherwood 1950, p. 153.
3 Butterfield 1941.
4 See for example Tony: https://code.soundsoftware.ac.uk/projects/tony.
5 Goodale 2011, pp. 1–2.
6 Roth 2004, pp. 29–30, 53.
7 Dr Abraham Wolfson to C.A. Lindbergh [all subsequent cited letters also to C.A. Lindbergh unless otherwise specified], 15 September 1941, Box 438, C.A. Lindbergh papers, Yale University Library [hereafter L-Y].
8 Brill 1927, p. 24.
9 E. C. McIntyre, 12 September 1941, Box 425, L-Y.
10 Mr and Mrs Wesley L. Johnson, 24 September 1941, Box 428, L-Y.
11 Lee Williams, 10 February 1941, Box 11, L-Y.
12 Mrs A. E. Roesel, 12 September 1941, Box 424, L-Y.
13 Ella Winkenwerden, 16 September 1941, Box 424, L-Y.
14 Mrs E. Hessel, 14 September 1941, Box 428, L-Y.
15 Violet Noll, 19 August 1941, Box 423, L-Y.
16 Brill 1927, p. 132; on Lindbergh's continuing appeal to women see Frost 2011, p. 73.
17 *Santa Ana Register* 1939.
18 *Moberly Monitor-Index* 1939.
19 Berg 1998, p. 397.
20 Levine and Levine 2002, p. 398.
21 *Chicago Tribune* 1940b; Lindbergh 1980, p. 134.
22 Lindbergh 1970, p. 375.
23 Speech, 19 October 1940, Box 202, L-Y.
24 *Franklin News Herald* 1940.
25 Lindbergh 1970, 411.
26 Schivelbusch 2006, p. 70.
27 Birdsall 2012, p. 112.

28 *Seattle Daily Times*, 24 April 1941.
29 Lindbergh 1980, p. 178.
30 The period of greatest Jewish migration to the United States had been some-what earlier, 1880–1924, when two and a half million Jews mostly from eastern Europe arrived in the country; only a quarter of a million mainly Jewish refugees entered the United States between 1934 and 1941. Jews in 1940 comprised less than 4 per cent of the total U.S. population. They were, it is true, concentrated in particular places—in cities over 100,000 the Jewish population was nearly 11 per cent and the New York metropolitan area housed over two million of the nation's estimated 4.7 million Jews. But in most ways the narratives and asser-tions of anti-Semitism had little to do with actual Jewish people, and cannot be explained by their presence.
31 Medhurst 2003, pp. 5–6.
32 Blakey 1992, pp. 13–14.
33 Lindbergh 1940, p. 43.
34 Lindbergh to Major John E. Kelly, 12 September 1940, Box 38, L-Y.
35 Lindbergh to William Castle, 18 October 1940, Box 38, L-Y.
36 Lindbergh to Colonel McCormick, 15 April 1941, Box 38, L-Y.
37 Turner 1941.
38 Simmons 1941.
39 Speech in St Louis, 3 May 1941, Box 202, L-Y.
40 Speech in Fort Wayne, 3 October 1941, Box 202, L-Y.
41 Woods 2014, pp. 6–7.
42 Butterfield 1941, p. 68.
43 DR 1939.
44 Lindbergh 1970, p. 404.
45 Schneider 1989, p. 12.
46 Stearns 1994, p. 193.
47 Louis Ludlow to CAL, 5 August 1940, Box 20, L-Y.
48 Mary McGowan, 25 September 1941, Box 427, L-Y.
49 Perle P. Fallon, 30 August 1941, Box 423, L-Y.
50 L. James Nagy, 15 September 1941, Box 424, L-Y.
51 Elizabeth Rabe Werkmeister, 9 August 1941, Box 423, L-Y.
52 Catherine Lewis, 12 August 1941, Box 423, L-Y.
53 *Chicago Tribune* 1940a.
54 Lindbergh 1970, p. 494.
55 Lindbergh 1980, p. 55.
56 *New York Times* 1941.
57 Lindbergh 1980, p. 190.
58 Two annotated drafts with 'A.M.L.' handwritten on top left-hand corner of first page, Box 202, L-Y.
59 *New York Times* 1939; Lindbergh 1980, p. 53.
60 ibid., p. 181.
61 *Chicago Tribune* 1941; Lindbergh 1980, p. 195.
62 Lindbergh 1970, p. 435.
63 One partial exception is Cole 1974.
64 Lindbergh 1980, p. 224.
65 Agnes Hovde, 10 August 1941, Box 423, L-Y. The two Agnes Hovdes in that region had either Norwegian parents or grandparents—close family ties to an occupied European nation only made this self-dedication to Lindbergh's leader-ship and his voice more striking.
66 My thanks to Henry Reese for drawing my attention to the apparent paradox of an unemotional patriotism.

References

Berg, S., 1998. *Lindbergh*, New York: G. P. Putnam.

Birdsall, C., 2012. *Nazi soundscapes: Sound technology and urban space in Germany, 1933–1945*, Amsterdam: Amsterdam University Press.

Blakey, G. T., 1992. 'Willkie as a Hoosier', in J.H. Madison, ed., *Wendell Willkie: Hoosier internationalist*, Bloomington, IN: Indiana University Press, pp. 3–21.

Brill, A. A., 1927. 'Why we mob heroes', *Popular Science Monthly*, September.

Butterfield, R., 1941. 'Lindbergh', *Life*, 11 August, p. 67.

Chicago Tribune, 1940a. 'Doctors advise war worriers to take it easy', *Chicago Tribune*, 18 June, p. 6.

Chicago Tribune, 1940b. 'Text of address by Lindbergh', *Chicago Tribune*, 5 August, p. 4.

Chicago Tribune, 1941. 'Willkie wants U.S. to set up Iceland bases', *Chicago Tribune*, 7 June, p. 3.

Cole, W. S., 1974. *Charles A. Lindbergh and the battle against American intervention in World War II*, New York: Harcourt Brace Jovanovich.

DR, 1939, 'Books, mostly', *Kokomo Tribune*, 16 September, p. 5.

Franklin News Herald, 1940. 'Lindbergh assailed by Wallace at Erie', *Franklin News Herald*, 16 October, p. 6.

Frost, J., 2011. *Hedda Hopper's Hollywood: Celebrity gossip and American conservatism*, New York: NYU Press.

Goodale, G., 2011. *Sonic persuasion: Reading sound in the recorded age*, Urbana: University of Illinois Press.

Levine, L. W. and Levine, C. R., 2002. *The people and the president: America's conversation with FDR*, Boston: Beacon Press.

Lindbergh, A. M., 1980. *War within and without: Diaries and letters 1939–1944*, New York: Harcourt Brace.

Lindbergh, C. A.: C. A. Lindbergh papers, Yale University Library (L-Y).

Lindbergh, C. A., 1940. 'Strength and peace', *Vital speeches of the day*, VII, 2, 1 November.

Lindbergh, C. A., 1970. *The wartime journals of Charles A. Lindbergh*, New York: Harcourt Brace Jovanovich.

Medhurst, M. J., 2003. 'Presidential speech writing: Ten myths that plague modern scholarship', in K. Ritter, ed., *Presidential speechwriting: From the new deal to the Reagan revolution and beyond*, College Station, TX: Texas A & M University Press, pp. 3–19.

Moberly Monitor-Index, 1939. 'Lindy says stay out', *Moberly Monitor-Index*, 16 September, p. 1.

New York Times, 1939. 'Text of Lindbergh neutrality talk', *New York Times*, 16 September, p. 7.

New York Times, 1941. 'Text of the addresses by Lindbergh and Wheeler', *New York Times*, 24 May, p. 7.

Roth, P., 2004. *The plot against America*, London: Jonathan Cape.

Santa Ana Register, 1939. 'Col. Lindbergh enters fight', *Santa Ana Register*, 16 September, p. 2.

Schivelbusch, W., 2006. *Three new deals: Reflections on Roosevelt's America, Mussolini's Italy, and Hitler's Germany, 1933–1939*, New York: Picador.

Schneider, J. C., 1989. *Should America go to war? The debate over foreign policy in Chicago, 1939–1941*, Chapel Hill, NC: University of North Carolina Press.

Sherwood, R. E., 1950. *Roosevelt and Hopkins: An intimate history*, New York: Grosset and Dunlap.

Simmons, R. C., 1941. 'Still hearing freedom of the press', *Chicago Defender*, 26 April, p. 15.

Stearns, P.N., 1994. *American cool: Constructing a Twentieth-century emotional style*, New York: NYU Press.

Turner, T.G., 1941. 'Lindbergh still "bashful" ', *Los Angeles Times*, 20 June, p. 1.

Woods, M.E., 2014. *Emotional and sectional conflicts in the antebellum United States*, New York: Cambridge University Press.

5 Noisy Classrooms and the "Quiet Corner"

The Modern School, Sound and the Senses[1]

Kate Darian-Smith

In A.S. Byatt's novel *Babel Tower*, set in Britain during the revolutionary 1960s, a Ministry of Education committee visits a 'new and exciting' primary school in Leeds:

> It is wholly glass-walled, and is built in the shape of a star. It is wholly open-plan: the children gather in little, impromptu-looking groups, in one or another arm of the star, bringing with them their brightly coloured bean bags, their plastic stools, their little tables. They are grouped not by age, nor by subject, but by some sort of self-directed choice of activity. . . . Small people dart busily and loudly from open space to open space. . . . Someone is playing the recorder in one star-tip, and someone near her is banging a drum. . . . There is a lot of noise. It is on the whole purposive noise, shrill, variegated, busy but loud.[2]

While 'Star Primary School' is an imaginative site, Byatt's description evokes the radical architectural and pedagogical approaches to education that were prevalent in the educational reforms of the post-World War II decades, not only in Britain but in many developed countries, including Australia. These approaches led to schools that were designed on open-plan principles, embracing natural light as, to use Catherine Burke's term, 'a technology of the classroom' associated with reason and progress.[3] Many of the conventional and often stark classrooms in older school buildings were modified or replaced entirely by learning spaces that were open to an array of sensory stimulators deliberately chosen to encourage learning. In these new educational spaces, the senses themselves—of sight, sound, touch, taste and smell—were also embraced as tools for teachers and students alike in the instruction and reception of knowledge.

This chapter examines the sensory environment of the modern Australian school from the 1940s to the 1980s, a period notable for the rapid expansion of the national education sector and intense public discussion about the ideal environment and methods for teaching future citizens. Widespread social disruption and the halt in the provision of adequate school facilities during the 1930s Depression directed government and public attention to the importance of education. During World War II, federal planning for post-war society recognized education reform—from the upgrading of inadequate school buildings to revised curriculum, raising of the minimum leaving age and improved teacher training—as key to securing Australia's economic and vocational opportunities for subsequent generations.[4] Such aspirations were to be realized through substantial growth, especially in secondary education, in the post-war period. In Britain too, the 1944 *Butler Act* created free secondary education for all and a programme of school construction that promised to educate and nurture children so as to unleash their potential in a democratic and more equitable society.[5]

Although Australian educationalists were influenced by British and other international ideas about pedagogy and school design, the Australian school system remained distinctive in many ways, including the size of Catholic education, especially at the primary level. The states were responsible for funding school education, resulting in regional differences. For example, New South Wales supported large comprehensive high schools tied closely to the surrounding neighbourhood, while Victoria maintained a dual system of technical and academic high schools up to the early 1980s. The astounding growth of education provision in the post-war decades, however, was shared across Australia and was driven by wider national priorities. The launch in 1947 of a mass migration scheme to bolster economic growth led to the arrival of one million migrants over the next decade, including many from non-English speaking backgrounds. With a parallel rise in the annual birth rate, student demand for schools rose by almost 60 per cent between 1950 and 1960, and a further 30 per cent by 1970.[6] Students were also staying at school longer. The minimum leaving age was raised to 15 years in most states, and the proportion of students completing high school, well under 20 per cent in 1945, rose steadily.

These demographic factors meant that new schools had to be built quickly and cheaply to educate the children of thousands of families, many of whom were moving into new suburbs on the metropolitan fringes.[7] The majority of existing school buildings required repairs and extensions, and there was an ongoing shortage of qualified teachers, especially in rural areas. By the 1950s, overcrowding was common, with some classes of up to 60 students taught by just one teacher. Education departments installed portable 'demountable' classrooms, and even ex-military Nissen huts, on many school campuses as a temporary measure to meet the unprecedented demand.

School has been the universal experience for Australian children during the twentieth century, regardless of their family, economic or geographical circumstances. Those children who attended school in the immediate post-war decades experienced its institutional confinement in a multitude of physical and emotional ways, not least through the senses. Indeed, many adult memories of school have a sensory immediacy in their describing of and re-experiencing of the feelings tied to that very specific place of childhood.

The senses can also be a powerful mnemonic device releasing a 'trigger to remember'—so that the smell of chalk dust or the ring of a school bell may elicit adult recollections of a classroom or playground in vivid detail.[8]

In published memoirs and autobiographies, the school is a prominent site and the setting for childhood dramas of injustice or triumph that are recounted with an emotional intensity as the narrator gains awareness of his or her self as a social being. In their anthology of Australian school-days, Brenda Niall and Ian Britain point out that the first day of school is often remembered as the first conscious separation of the child from home and parents. This highly personalized moment can be situated within and confirmed by a wider historical context: 'The date, place, buildings, and uniforms are matters of verifiable record that the autobiographical self can retrieve and muse upon.'[9]

In her examination of the memories of an Australian childhood during the 1950s, Carla Pascoe identified three interlinked spaces: the neighbour-hood, the home and the school. School—attended by children from as young as four years old to their mid-teens, for five days a week, and for most weeks of the year—was characterized by greater external regulation than found in the neighbourhood or home, with discipline imposed by adults who had no family or emotional ties to the children in their care. Five decades later, Pascoe's interviewees recalled their school lives through embodied memories of place. These include the memories of everyday injuries, 'such as an ink pen jabbed into the hand, backside splinters from the classroom floor, and bloody knees sustained through falls on to the gravel,' as well as the intimate sensory details of 'warm milk, spilt blue ink, or chilly classrooms.'[10] The common presence of material objects in school memories such as these, Lisa Rasmussen argues, 'come to reflect the material, sensuous and relational dimensions of everyday school life.'[11] While objects are often the catalyst for narratives, they also function to keep adults either literally or figuratively 'in touch' with the past, and the events or feelings that make us 'who we are.'

In recent years, educational history around the world has paid increasing attention to the 'practice, meaning and culture of classrooms' across time.[12] This work has not only interrogated the spatial politics, architecture and materiality of the school,[13] but has also explored the sensory dimensions of educational experiences from perspectives of teacher and pupil. Ian Grosve-nor's useful mapping of some studies of schooling and the senses has shown how these range from the exploration of touch and its use in the education of the 'feebleminded' during the nineteenth century, to research arguing that efforts to create 'the perfect eye' in children were hampered by 'defective illumination, poor desk design, habitual writing exercises, needlework drill, unsuitable letter fonts, and poor quality writing,' which combined to cre-ate 'an environment deleterious to normal ocular development.'[14] Sight is perhaps the most explored sense in sensory histories of schooling. Ques-tions of pedagogy, design and social value were brought together in the progressive education of 'the eye' and the recognition of the importance of 'visual pleasure' in the planned environments and buildings of the modern school.[15] For instance, the responses of children to the display of artwork in the classroom can be linked to broader questions of citizenship and national identity.[16] Other work has turned to the smells and tastes associated with the school, with Burke examining how 'food and drink and the space in which they are served and consumed can become a site of contested desires, a space where authority and resistance are exercised.'[17]

The soundscapes of schooling have also attracted increased scholarly attention. Burke's and Grosvenor's analysis of 'The Hearing School' traces the networks of sound that defined the English school experience during the first half of the twentieth century. From pedagogical practices and school routines to the design of educational spaces, the mechanics of auditory control and the introduction of technologies of learning, they demonstrate that the sense of sound is deeply embedded in 'the daily elaboration of modern schooling.'[18] This chapter's investigation of schooling in Australia in an overlapping historical period outlines several complimentary developments in the soundscape of the modern school. The introduction of 'child-centred' pedagogies, the increasing application of audio-visual technologies in the teaching and learning process and changes in the curriculum of music education were all factors that would have an impact on the auditory experience of school. By the late 1960s, emerging concerns were being expressed by education departments and professional teacher associations about the disruptive and unproductive 'noise' now heard in schools. Rising auditory levels were linked to student inattention and potential learning difficulties, while the 'noisy classroom' was perceived as the site of heightened psychological stress for teachers.

The Modern Australian School

In 1954, American folklorist Dorothy Howard visited Australia on a Fulbright fellowship to document the folklore of Australian children, travelling for almost a year across the continent to observe the games that children played in primary school playgrounds during lunch and recess breaks. She found that 'below the eye level' of adults, children played hundreds of traditional games. The sounds of the playground were loud and varied. Children's chants accompanied rhyming, clapping and skipping games, balls were bounced off walls, cries and yells were associated with cricket, football and chasey, and all this was underpinned by the hum of children's conversation unfettered by adult supervision.[19] Although the majority of school playgrounds in the 1950s were physically cramped and inadequate, regulated by teachers on 'yard duty' and divided into gender- and age-specific zones, nonetheless ideas about the importance of 'free' play to individual child development were gaining wider circulation.

Progressive ideas were also transforming philosophies of school design. The crisis of mid-twentieth-century schooling in Australia was partially addressed by a boom in the construction of new schools during the 1950s to meet the escalating demand. State education departments developed standard architectural models that were relatively inexpensive to erect and were adaptable to local climate conditions and the availability of building materials. School design was underpinned by the belief that well-planned educational environments would instil ideas of responsible and participatory citizenship among Australian children, including those of recently arrived migrants. Inside the classroom, curriculum and pedagogy were increasingly shaped by new understandings of child psychology and parenting that recognized the unique potential of the individual child and prescribed ways in which this could be supported.[20]

The educational and popular focus on the development of the 'whole child' had major implications for teaching practice in Australia. A 'child-centred' education—as seen in Byatt's description of the Star Primary School—removed the teacher as the sole authority figure, and instead emphasized the self-direction of children in their learning. A publication by the Victorian Education Department on 'The Primary School' summed up this philosophy by stating that each child must be allowed 'optimum development,' with the 'development of the individual person at the heart of the education process.'[21] Pedagogical methodologies were thus aimed at catering for individual capacities and inclinations across the class cohort. The teacher was no longer cast as the authority figure located at the front of the class, dispensing information to a silent mass of desk-bound children, but the facilitator of an interactive and verbal group of students who were free to move around the room. Teacher training, which had previously focused on the modulation and level of the teacher's own voice, now switched to strategies that would encourage students to speak for themselves.

The shift towards progressive education was to alter the soundscape, particularly in relation to the spoken word, within the classroom. By the mid-twentieth century, educational research had established the importance of spoken language for cognitive learning and personal development. In a school setting, this was now linked to recognition of the significance of the talk that students conducted among themselves, especially in relation to grasping new concepts. To take one example, the prominent British study published in 1969, *Language, the Learner and the School*, argued that students' use of oral language could lead to new ways of thinking, feeling and representing reality.[22] Such research and its radical implications were widely discussed among Australian educationalists. Teachers—who were observed speaking for two-thirds of available class time—were asked, 'Do you talk too much?' From the perspective of the spoken word, the aural environment of the classroom could be divided into 'teacher talk' or 'student talk,' with periods of 'silence or confusion' in-between.[23] Nonetheless, annual reports on student achievement sent to parents indicated (and still do) that many teachers complained about 'too much talking in class' by high-spirited students and saw 'excessive' student 'noise' as a matter for disciplinary action.

The level of optimum sound in schools was to become a topic of public discussion and recognized the differences between an optimum auditory environment for young children and one appropriate to teenagers. The organization of high school content and timetabling into segmented blocks of subjects throughout the day contrasted with primary schools where students spent the day in one classroom space. There were, however, many common auditory elements between primary and secondary schools. The influential educationalist Henry Schoenheimer, who through his regular commentary in *The Australian* newspaper between 1965 and 1975 brought ideas about child-centred schooling into the popular domain, advised parents to use their 'eyes and ears' in assessing a 'good school' for their children. Listening in the corridor to the sounds of a school was an important test:

> If the classrooms are unexceptionally quiet and orderly: if the only sounds you hear are those of the teacher teaching or the radio, TV, film or tape recorder dispensing knowledge—that's a bad school. . . . In a good classroom, there is frequently the inescapable and easily audible hum and constant chatter of groups of people busy on common tasks in small spaces. . . . All good classrooms spend a lot of quiet time in thought, especially in the senior school. But a silent school is a bad school, just the same. And in a school of twenty or thirty classes, there should be singing and music and poetry and drama at four or five of them at any time of day.[24]

Schoenheimer's advice on 'good' and 'bad' educational soundscapes was couched in terms of student creativity and productivity, and of the need for active student participation in the acquisition of skills and knowledge. The ratio between teacher and student talk had to be carefully calibrated, with 'quiet time' seen as a positive step in learning rather than a punitive check by the teacher on student oral expression or a time of 'confusion.' This issue of balance between the 'silent' and the 'noisy' classroom was to be one of the most persistent debates about school environments and the role of the teacher that arose in the post-war decades.

Media Technologies in the Classroom

The aural and visual stimulation provided by the new media technologies of film, radio and television were to change the learning environment of the classroom during the twentieth century, creating new sensory experiences of sight and sound. In 1924, the first Australian radio broadcast designed specifically for classroom use went to air in Sydney. By July 1929, the private Australian Broadcasting Company began transmitting a daily 'Education Hour.' The first 15-minute segment to be aired addressed 'Common errors in pronunciation', a subject particularly suited to a radio talk.[25] The Australian Broadcasting Company soon abandoned educational broadcasting, citing the conservatism of teachers in its limited take-up, although the absence of government funding for radio receivers in schools was a more likely explanation.[26] In the 1930s, according to historian K. S. Inglis, there was 'still scepticism and suspicion within Education Departments about the use of radio in the classroom.' When the national broadcaster, the Australian Broadcasting Commission (ABC), was formed in 1932, it took over much of the responsibility for educational broadcasting. The ABC gave assurances to teachers and education departments that radio 'must endeavour to supplement, so far as possible, and never to supplant, the work of the teacher in the school.'[27]

Early educational radio programmes on the ABC included 'Adventures in Music,' 'The World We Live In' and 'Health and Hygiene.'[28] These took the form of the 'straight radio lesson' or the 'radio talk,' with a local presenter reading a lecture with no other aural content, although music broadcasts were a notable exception. From a pedagogical perspective, educational radio talks maintained the authority of the teacher, or the disembodied voice of the substitute adult expert, and emphasized the power of the spoken word—now employed, through transmission, in mass education. Children were required to sit quietly in the classroom and listen carefully to a radio set, usually located on the teacher's desk at the front of the room, just as they were expected to listen to their teacher.

During the 1930s, the majority of educational segments were produced quickly, often as part of the trading of presenters who were inexperienced in the intricacies of broadcasting.[29] After World War II, educational services expanded and production was professionalized. The ABC began working closely with staff from state education departments to ensure that the content of educational programmes was related to the curriculum and could be readily incorporated into daily lesson plans.[30] Specialist periodicals and a series of 'Notes' for teachers were published to support the use of the broadcasts.[31] By the mid-1950s, 88 per cent of all Australian schools had access to radio sets, and in many classrooms there was regular, even daily, use of the ABC's programmes for schools.

Radio technology also provided new educational opportunities for those children living in remote areas of Australia who studied via correspondence and had no direct contact with a teacher or fellow pupils. In 1951, the School of the Air was launched at the Flying Doctor Base in Alice Springs, in the Northern Territory, and other branches were soon operating around the country. Initially, one-way radio lessons were transmitted to isolated students, but this evolved into a two-way exchange with a question and answer session at the end of each broadcast. This was an international innovation and swiftly transformed distance education in Australia, and later in other parts of the world. The first book-length study of School of the Air, published in 1971, was evocatively titled *Out of the Silence*. A more recent evaluation has claimed that the School's students experienced a unique sonic education, gaining familiarity with 'ring modulation, phasing, low frequency oscillation, distortion, compression, white noise, pink noise and the like, not to mention the use of microphone as vocal extension.'[32]

In the mainstream classroom, visual aids such as posters were becoming more common too, either used independently or as an accompaniment to aural material. The philanthropic Carnegie Corporation of New York, which founded the Australian Council for Educational Research and instigated the Munn-Pitt review of Australian libraries in the 1930s, allocated over £10,000 for teaching kits on art and music for Australian secondary schools. The art kits included books and over 900 photographs and colour reproductions as an exercise in the shaping of middlebrow cultural taste, but their circulation was very limited.[33] It was the humble 'filmstrip,' a sequence of images in a spooled roll of 35 mm positive film, that was to dominate visual media in the classroom from the 1940s to the 1980s. The filmstrips and projector were relatively inexpensive, and while some filmstrips were imported, content could also be produced locally by education departments and tied in with primary and secondary school curriculum. Teachers were often given notes or a prepared 'script' to accompany each slide. While the showing of a filmstrip remained teacher-centred, it did require a darkened space and so altered the classroom. By 1969, the ABC began recording some radio programmes on tape for school use and developed Radiovision kits, comprising written materials and a filmstrip designed to be screened alongside an accompanying radio broadcast.[34]

Showing films to children was a complicated affair, given the need for an expensive projector and the skills to operate it. From the 1940s, state education departments had teams of roving projectionists who visited schools on request.[35] Television was less complex. After 1956, when television was introduced into Australia, it was incorporated into the sensory learning environment of schools. The ABC produced its first educational television for secondary schools in 1957, and for primary schools in 1958. Education departments favoured science and mathematics content for older students, and 'Behind the News,' a magazine-style programme still aired today, proved particularly popular. By mid-1965, about 25 per cent of Australia's 10,000 schools had access to television, with this proportion rising to 84 per cent by 1972.[36] Most schools had only one television set (compared to an average of five radios), and this was usually mounted on a trolley so it could be easily moved. In general, radio and television broadcasts were used more frequently in the primary level classroom, as there were timetable restraints in secondary schools. The introduction of video recorders in the 1970s gave teachers greater flexibility, removing the tyranny of fixed transmission times.

Writing in *The Educational Magazine* in 1977, former teacher and education journalist Geoffrey Maslen commented that in most schools, televisions were 'fixed permanently to Channel 2 [ABC] in case some deviant falls prey to the temptation to watch something entertaining.' Closed-circuit television equipment and videotape-recording facilities were becoming more common. In most cases, videos were used to record the ABC's educational broadcasts, or select programmes on commercial stations that aired in the evening, so these could be run at a later date. Maslen and others were critical of the often stagnant teacher use of audio-visual technologies:

> TV in school is merely a sophisticated classroom aid—a glorified mobile blackboard. Only rarely do students get to use such equipment, to create their own TV productions, to explore for themselves the medium's peculiar problems and potential. Instead of learning to confront the artificial world of TV, of studying its moulding power, students learn only passivity. The inert environment of the chalk-and-talk classroom remains unchanged by the talking head on the TV screen.[37]

Audio equipment in the classroom extended to language teaching. Most high schools were equipped with dedicated alcoves and 'laboratories' where students could develop their oral and aural skills in a 'foreign' language or, in the case of migrants from non-English speaking backgrounds, in English. Tape recording was also useful in assisting with pronunciation and expression. Annual reporting on the teaching and learning standards in primary schools by state education departments included comments on the aural capacity of students. For instance, in October 1967 the District Inspector of Education wrote that at the rural Glengarry West Primary, in the Gippsland region of Victoria: 'The pupils speak freely and interestedly on a wide range of subjects. However many speech defects remain noticeable. Use of the tape recorder to promote self-listening could do much to improve this aspect of the work.'[38] At high schools, speeches on topics of national interest remained within the curriculum, and the annual practice of Speech Night, where teachers and students deliver a public address, continues to this day.

Listening to and Making Music

Sound technologies were especially suited to the teaching of music and oral expression in schools. Aural components of the school curriculum were valued from the nineteenth century as a disciplinary and moral force, with the words of patriotic songs or poems rousing a sense of national and imperial loyalty in children. At the turn of the twentieth century, schools in Australia were musical places. Snatches of choral music could be heard throughout the school day, often starting with a school assembly, and students were familiar with the bodily experience of singing in groups, including for public performances. Instruction in singing 'by ear' commenced in infant classes and was continued with increasing levels of complexity through every year level. The assembly of choirs of up to 10,000 school-children in the state capitals of Melbourne or Sydney to celebrate major events, such as the inauguration of Federation in 1901 or the visit of the Prince of Wales to Sydney in 1920, attests to the widespread participation of children in mass singing.[39]

By the 1920s, the use of gramophones and radio broadcasts in the class-room launched the study of musical appreciation—or music listening. In 1929, well-known conductor Bernard Heinze organized a concert at the Melbourne Town Hall for school audiences. Similar educational perfor-mances followed in other states and were to become an annual tradition. State education departments established free lending libraries of gramo-phone records for school use, and these were highly subscribed to. The introduction in schools of the 'Music through Movement' approach, based on Dalcroze eurhythmics, was supported by ABC radio broadcasts from 1938.[40]

Nonetheless, by the 1930s school inspectors in NSW raised concerns about music education, linking the reproduction of music through the radio or gramophone to an increase in teachers with no instrumental or technical knowledge of music.[41] One solution enacted in Sydney metro-politan schools during the 1940s and 1950s was the organization of com-bined primary school choral concerts.[42] At the national level, there was a concerted shift in music education to encompass not only vocal work, but also the playing of instruments. This included the establishment of percussion bands in primary schools during the inter-war years, a measure seen to aid young children with the development of mental concentration, discipline and participatory citizenship. One of the barriers to percussion bands was that some teachers found the music to be unattractive, and the 'noise' could disturb other classes. This was less so for the introduction of flute and recorder bands for older children, following the success in Brit-ain of schools-based recorder playing. By 1941, for instance, there were over 70 school flute bands in Sydney, and another 30 in country towns throughout NSW, although numbers were to decline in the following decades.[43] The inexpensive and portable recorder was even more popular, with more than 10,000 recorders sold to primary children in NSW alone between 1945 and 1952.[44]

With choral and instrumental music part of the everyday aural envi-ronment of schools in the mid-twentieth century, new organizational and spatial approaches to children's listening were introduced. The 1952 NSW Curriculum for Primary Schools advised that 'in listening lessons an atmo-sphere of quiet relaxation is essential [and] the children should be grouped informally and allowed to sit in a relaxed, comfortable position.'[45] A decade later, creative music making was gaining prominence, influenced by devel-opments in England and discussions at an influential UNESCO Conference on School Music held in Sydney in 1967.[46]

The new emphasis on making music at both primary and secondary schools meant that class singing slowly lost its dominance, and music theory was in decline. A report on music in Victorian primary schools in 1977, published in the teachers' journal *The Educational Magazine*, noted that the 'common goal' of primary music teachers was 'to give as many children as possible the satisfaction of musical expression, and through participatory and listening activities to inculcate a love of music and some glimmering of musical discrimination.' With no set curriculum in primary music education, the Victorian Education Department provided resource materials for class activities, designed for the 'non-specialist' teacher. These included ABC broadcasts of 'traditional' songs, with 'new songs' available in the department's own songbooks and tapes. Yet, the report commented, 'despite the existence of plentiful material, what schools can achieve varies with the facilities and teacher-talent available.'[47] Another article in the journal criticized the 'lack of real commitment' to music as a component of 'total education,' fearing that the 'fragmentation and compartmentalisation of musical practice' had led to an outdated and piecemeal approach. It concluded that 'children should be educated for their entire life and not simply for their school years. To develop the full potential of children, educators must take a new look at music, and music education.'[48]

Noisy Classrooms

As progressive approaches to teaching were adopted throughout Australian schools, these were accompanied by interventions in school design that recognized a shift away from the dominance of teacher-directed learning. In most conventional primary school classrooms, a 'quiet corner' was introduced to encourage children to read, draw or initiate other activities that required little or no talking—and thus take time 'out' from the sensory stimulation of a modern active class, where different tasks were conducted simultaneously. As part of a wider international trend, by the early 1970s the open-plan classroom was adopted in many Australian schools as an architectural solution to the emerging demand for collaborative modes of teaching and learning.[49] In 1972, an observer of 72 children in Grade 6 undertaking 'integrated studies' in an open-plan classroom at Thomastown Primary School, on the northern fringe of Melbourne, commented that:

> They open the folding doors and move freely about the building to work on topics that they have chosen for themselves. Discussion is encouraged and there is a free exchange of ideas between individuals and groups *but the noise level is not unduly high* [my italics]. The plentiful audio-visual equipment is used by the children, not the teachers.[50]

Nonetheless, such fluid spatial arrangements in combination with 'freer' educational methods led to a marked increase in the concerns raised in

Figure 5.1 'Open Space Primary Schools,' *The Educational Magazine* 29, no. 4 (1972), pp. 13–15. State Library of Victoria.

educational department reports and teachers' journals about the auditory levels of the classroom. Winifred Jeavons, a teacher researching for a masters thesis on noise levels in schools, wrote that 'there are probably few teachers who have not experienced the nuisance quality of noise.' For Jeavons, different types of noise disrupted the classroom and student learning. There was noise generated by external traffic or the use of air-conditioners, but also the student-led noise of the 'unmanageable class' or the teacher-led noise of the music lesson. Jeavons pointed to 'a design-to-function mismatch' in the 1970s school. The 'practice of siting long lines of lockers along highly reverberant corridors' or the replacement of old-fashioned screwed-down furniture with portable tables and chairs which scraped against floorboards served to amplify noise levels. Moreover, school buildings and classrooms, some dating back to the Victorian era, were sonically inadequate when it came to modern educational approaches:

> Schools designed to restrain children in their desks and in their rooms for blocks of time, with corridors empty and silent except at change-over times, simply don't function as well with freer teaching and time-tabling procedures. If you want students to move freely . . . then you must design for it, and not assume that good intentions are an adequate replacement for good acoustic control.

Jeavons believed that noise could be curtailed by the installation of carpets in corridors or classrooms, furniture that had sound-absorbent impact

Figure 5.2 Joe Kolker, 'A Quiet Place: A Modern Education Parable,' *The Educational Magazine* 33, no. 6 (1976), pp. 18–20. Curriculum Branch, Education Department of Victoria. State Library of Victoria.

points and lockers that would not rattle or were isolated from teaching areas.[51]

While concerns about unacceptable noise were generally couched in terms of the interests of students, the impact of the new intensities of sound on the vocational experiences of teachers also created difficulties. This was illustrated in a short story entitled 'A Quiet Place: A Modern Education Parable,' published in 1976. Aimed at readership among the education profession, this was a fictional account of a shy young teacher who sees employment in a school library as his destiny, as it is a quiet place. But he is forced by the school authorities to introduce audio-visual equipment, and the structured timetable for student use of the library is replaced by more open and fluid access. These measures transform the library from 'an oasis of peace' to being like 'Bourke St' [Melbourne's busiest shopping street] and becoming 'bedlam,' where children come and go at will and books are no longer read. The distressed teacher eventually gives up teaching altogether, and finds employment at a municipal library 'where the rule of silence still prevails.'[52]

Conclusion

As this chapter has shown, the post-war expansion of schools in Australia, and the accompanying adoption of progressive educational practices, resulted in a new awareness of auditory levels and controls within modern educational environments. The arising ambiguities and ambivalences about sound and 'noise' in relation to schooling were embedded within the entire educational system. Sound was a crucial component in the architecture of school buildings, the internal organization of classrooms and the provision of technological infrastructure. Acknowledgement of the aural expression and reception of students was central to educational policies across teaching practice and curriculum implementation, and was evident in the everyday rhetoric about 'good' schools, the language of the report card and the value placed on forms of student and teacher behaviours in relation to sound.

This attention to noise in school environments has never since abated, although auditory expectations and ongoing discussions about the dynamics between sound and classroom design, pedagogy and learning have shifted. By 1980, complaints that noise levels in Australian schools were having an impact on children's ability to learn were becoming more strident. In 1983, it was reported that the acoustic level in the average classroom was around 70 decibels, so that teachers and students had to raise their voices to 'communicate intelligibly at a maximum distance of approximately two metres.' In search of a solution, one teacher pasted cardboard egg cartons onto the entire ceiling of her classroom, with effective results in lowering the auditory level.[53]

Research has now confirmed that long-term exposure to excessive environmental noise can result in temporary or permanent hearing loss, with the incidence in Australia on the rise. A recent report on national educational achievement claimed that 'Australian students report high levels of noise and disruption in their classroom and at rates worse than the US or Britain,' a factor linked to declining standards of literacy and numeracy across the nation.[54] An opinion piece recently warned that while it is known that noise interferes with 'attention, concentration and thinking,' 'the precise impacts of noise on critical child learning capacities haven't been measured,' and we should be 'especially worried about the health and well-being of children in the longer term.'[55]

At the same time, 'agile learning spaces' have been re-discovered, with some Australian schools now introducing groupings of up to 120 children and several teachers in a single large space. One school principal complained that conventional classrooms were 'getting in the way' of the 'independent learning' of students, and of the efforts of teachers to 'help the children discover things for themselves.'[56] These ideas are reminiscent of experiments in open-plan classrooms dating from the 1960s, with many ultimately deemed to be a failure because children were too distracted by multiple activities and sensory stimulation. Today's educators point to important differences in technology, architecture and pedagogy that mean the old mistakes will not be repeated. A *Sydney Morning Herald* journalist, visiting a 'mega classroom' of 197 students at St Monica's Primary in the Sydney suburb of North Parramatta, saw and heard 'different-sized groups at a range of tasks at the same time, seemingly without disturbing each other. . . . Certainly it is noisier than a traditional classroom but the children did not appear to be distracted.'[57]

International research suggests otherwise. The twenty-first century move to more open-plan schools in Norway to cope with financial constraints and fluctuating student numbers has led to 'more noise, less concentration, and the practice of ability grouping that is pushing the limits of what is permissible under current regulations.'[58] An earlier report into 'classroom design for good hearing' by the Noise Pollution Clearinghouse, an American non-profit organization, pointed to how children in open-plan classrooms 'could hear the teacher of an adjacent class more clearly than their own teacher.'[59]

Today's debates about noise in the modern classroom thus continue a longer historical narrative around education and sound, and the sensory world of schooling more broadly. The historical experiences of the post-World War II school in Australia still exist within living memory, and will continue to be part of a remembered childhood by those generations of Australians who are now moving from middle age to retirement in their life cycle. The relational concepts of 'noise' and 'quiet' in educational environments that emerged from the 1950s point to the rapidity in which a combination of pedagogical and technological innovation changed the soundscape of learning in ways that were irreversible and transformative—and with reverberations still current in contemporary educational discourse.

Notes

1 Research for this article was conducted with support from a grant from the Australian Research Council (DP110100505), and I am grateful for the insights of Dr Sianan Healy.
2 Byatt 1996, p. 172.
3 Burke 2005b.
4 Macintyre 2015, pp. 213–18.
5 Saint 1987, p. 35.
6 McQueen 2004, p. 220.
7 Campbell and Proctor 2014, pp. 178–210; Healy and Darian-Smith 2015.
8 Hamilton 2011.
9 Niall and Britain 1998, p. xiii.
10 Pascoe 2011, p. 193.
11 Rasmussen 2012, p. 115. See also Dane, Earle and van Ruiten 2011.
12 See for an early intervention on classroom culture, Grosvenor, Lawn and Rousmaniere 1999.
13 See Burke and Grosvenor 2008.
14 Grosvenor 2012, pp. 678–79.
15 Burke 2005b; Burke 2010.
16 Grosvenor 2005; Grosvenor 2011.
17 Burke 2005a.
18 Burke and Grosvenor 2011, p. 337.
19 See Darian-Smith and Factor 2005.
20 See Healy and Darian-Smith 2015.
21 Educational Facilities Research Laboratory 1971.
22 Barnes, Britton and Rosen 1969.
23 Anderson 1970.

24 Schoenheimer 1980, pp. 132–33.
25 Lapsley 1998, p. 5.
26 Inglis 1983, p. 15.
27 Quoted in Inglis 1983, p. 32.
28 Inglis 1983, p. 57.
29 Lapsley 1998, pp. 8–9.
30 Inglis 1983, p. 167.
31 See, for instance, the periodical Australian Broadcasting Commission, Victorian Branch, 1945–58.
32 Ashton 1978; Australian Broadcasting Corporation n.d.
33 *Sydney Morning Herald*, 19 April 1934, p. 3; *Sydney Morning Herald*, 11 August 1934, p. 10.
34 Nichols and Lewi 2017.
35 ibid.
36 Inglis 1983, p. 209; Perlgut 2014.
37 Maslen 1977, p. 33.
38 Report on Glenferrie West Primary School, 13 October 1967, Public Records Office of Victoria 1951–71.
39 Chaseling and Boyd 2014, pp. 46–47.
40 Stevens 1997; Stevens n.d.
41 Chaseling and Boyd 2014, p. 50.
42 ibid., p. 53.
43 ibid., pp. 55–57.
44 ibid., p. 57.
45 Quoted in Chaseling and Boyd 2014, p. 54.
46 Stevens n.d.
47 G. S. 1977.
48 McMahon 1977.
49 Cameron 2017.
50 Bullen 1972.
51 Jeavons 1976.
52 Kolker 1976.
53 Petropulos 1983.
54 Tovey 2013.
55 Prior 2013.
56 Overington 2011.
57 Stevenson 2011.
58 Erikson 2014.
59 Wetherill 2002.

References

Anderson, M. C., 1970. 'Do you talk too much? An analysis of talking in our classroom', *The Educational Magazine*, 27, no. 9, pp. 432–36.
Ashton, J., 1978. *Out of the silence*, Adelaide: Investigator Press. Reissued as *The school of the air*, Adelaide: Rigby.
Australian Broadcasting Commission, Victorian Branch, 1945–58. *ABC school broadcasts, secondary schools*, Melbourne: ABC.
Australian Broadcasting Corporation, n.d. 'Australia Adlib', Available at http://www.abc.net.au/arts/adlib/stories/s857521.htm, Accessed 2 September 2015.
Barnes, D., Britton, J. and Rosen, H., 1969. *Language, the learner and the school*, London: Penguin.

Bullen, P., 1972. 'Open space primary schools', *The Educational Magazine*, 29, no. 4, pp. 13–16.

Burke, C., 2005a.'Contested desires: The edible landscape of school', *Paedagogica Historica: International Journal of the History of Education*, 41, pp. 571–87.

Burke, C., 2005b. 'Light: Metaphor and materiality in the history of schooling', in M. Lawn and I. Grosvenor, eds., *Materialities of schooling: Design—technology—objects—routines*, Oxford: Symposium Books, pp. 125–44.

Burke, C., 2010. 'About looking: Vision, transformation, and the education of the eye in discourses of school renewal past and present', *British Educational Research Journal*, 36, no. 1, February, pp. 65–82.

Burke, C. and Grosvenor, I., 2008. *School*, London: Reaktion Books.

Burke, C. and Grosvenor, I., 2011. 'The hearing school: An exploration of sound and listening in the modern school', *Paedagogica Historica: International Journal of the History of Education*, 47, no. 3, pp. 323–40.

Byatt, A. S., 1996. *Babel tower*, London: Chatto & Windus.

Cameron, L., 2017. 'Open shut them: Open classrooms in Australian schools, 1967–1983', in K. Darian-Smith and J. Willis, eds., *Designing schools: Place, space and pedagogy*, London and New York: Routledge, pp. 113–31.

Campbell, C. and Proctor, H., 2014. *A history of Australian schooling*, Sydney: Allen & Unwin.

Chaseling, M. and Boyd, W. E., 2014. 'The decline and revival of music education in New South Wales schools, 1920–1956', *Australian Journal of Music Education*, 2, pp. 46–61.

Dane, J., Earle, S. -J. and van Ruiten, T., 2011. 'The material classroom', in S. Braster, I. Grosvenor and M. del Mar del Pozo Andres, eds., *The black box of schooling: A cultural history of the classroom*, Brussels: Peter Lang, pp. 263–76.

Darian-Smith, K. and Factor, J., eds., 2005. *Child's play: Dorothy Howard and the folklore of Australian children*, Melbourne: Museum Victoria.

Educational Facilities Research Laboratory, 1971. *Facilities for arts crafts rooms in primary schools*, Melbourne: Department of Education. In Public Records Office of Victoria, 'Details of planning including minor adjustments (BG37B)', VPRS 8819/P/0001, Unit 7,

Erikson, J., 2014. 'Children bear the brunt of open-plan schools', *Science Nordic*, 17 January. Available at http://sciencenordic.com/children-bear-brunt-open-plan-schools. Accessed 15 July 2014.

Grosvenor, I., 2005. ' "The art of seeing": Promoting design in education in 1930s England', *Paedagogica Historica: International Journal of the History of Education*, 41, no. 4–5, pp. 507–34.

Grosvenor, I., 2011. ' "To act on the minds of the children": Paintings into schools and English education', in S. Braster, I. Grosvenor and M. del Mar del Pozo Andres, eds., *The black box of schooling: A cultural history of the classroom*, Brussels: Peter Lang, pp. 39–57.

Grosvenor, I., 2012. 'Back to the future or towards a sensory history of schooling', *History of Education: Journal of the History of Education Society*, 41, no. 5, pp. 675–87.

Grosvenor, I., Lawn, M. and Rousmaniere, K., eds., 1999. *Silences and images: The social history of the classroom*, New York: Peter Lang.

G.S., 1977. 'Fabric or frill? Music in the primary schools and the role of the district adviser', *The Educational Magazine*, 34, no. 1, p. 19.

Hamilton, P., 2011. 'The Proust effect: Oral history and the senses', in D. A. Richie, ed., *The Oxford handbook of oral history*, New York: Oxford University Press, pp. 219–32. Reprinted in *Oxford handbooks online*.

Healy, S. and Darian-Smith, K., 2015. 'Educational spaces and the "whole child": A spatial history of school design, pedagogy and the modern Australian nation', *History Compass*, 13, no. 6, pp. 275–87.

Inglis, K. S., 1983. *This is the ABC: The Australian broadcasting commission, 1931–1983*, Melbourne: Melbourne University Press.

Jeavons, W., 1976. 'Sounding off', *The Educational Magazine*, 33, no. 6, pp. 32–33.

Kolker, J., 1976. 'A quiet place: A modern educational parable', *The Educational Magazine*, 33, no. 6, pp. 18–20.

Lapsley, T., 1998. 'The origins of educational broadcasting in Australia', *Australian Journal of Educational Technology*, 14, no. 1, Winter, pp. 1–24.

Macintyre, S., 2015. *Australia's boldest experiment: War and reconstruction in the 1940s*, Sydney: New South Books.

Maslen, G., 1977. 'Television and the uncaring teacher', *The Educational Magazine*, 34, no. 4, pp. 32–33.

McMahon, H., 1977. 'Music education is. . .', *The Educational Magazine*, 34, no. 1, pp. 6–7.

McQueen, H., 2004. *Social sketches of Australia*, Brisbane: University of Queensland Press.

Niall, B. and Britain, I., eds., 1998. *The Oxford book of Australian schooldays*, Melbourne: Oxford University Press.

Nichols, D. and Lewi, H., 2017. 'Bristling with opportunity: Audio-visual technology in Australian schools from the 1930s to the 1980s', in K. Darian-Smith and J. Willis, eds., *Designing schools: Place, space and pedagogy*, London and New York: Routledge, pp. 218–28.

Overington, C., 2011. 'Funky school', *The Australian*, 10 September. Available at http://www.theaustralian.com.au/news/features/funky-school/story-e6frg8h6-1226130668112?nk=bc5668bcec1e5c9c191c8506d93b434e. Accessed 15 July 2014.

Pascoe, Carla, 2011. *Spaces imagined, Places remembered: Childhood in 1950s Australia*, Newcastle upon Tyne: Cambridge Scholars Press.

Perlgut, D., 2014. 'Educational media', in B. Griffin-Foley, ed., *A companion to the Australian media*, Melbourne: Australian Scholarly Publishing, pp. 148–49.

Petropulo, A., 1983. 'Conditions for comfort', *The Educational Magazine*, 40, no. 5, pp. 15–17.

Prior, M., 2013. 'Wired by sound: The long-term impacts of constant noise', *The Conversation*, 19 February. Available at https://theconversation.com/wired-by-sound-the-long-term-impacts-of-constant-noise-11020. Accessed 10 September 2015.

Public Records Office of Victoria, 1951–71. 'School records, inspectors report book', VPRS 9351/P/0001, Unit 2.

Rasmussen, L. R., 2012. 'Touching materiality: Presenting the past of everyday school life', *Memory Studies*, 5, no. 2, April, pp. 114–30.

Saint, A., 1987. *Towards a social architecture: The role of school buildings in postwar England*, New Haven, NY: Yale University Press.

Schoenheimer, H., 1980. 'What to look for in a school—Any school', in L. Allwood, ed., *Schoenheimer on education*, Melbourne: Drummond Publishing, pp. 132–34.

Stevens, R., 1997. 'Music education in Australia', in W. Bebbington, ed., *The Oxford companion to Australian music*, Melbourne: OUP, pp. 396–99.

Stevens, R., n.d. 'History of music education in Australia', *Australian Music Education Information and Resources*. Available at http://australian-music-ed.info/History/index.html. Accessed 15 July 2014.

Stevenson, A., 2011. 'All in together—197 students in one room', *The Sydney Morning Herald*, 6 June. Available at http://www.smh.com.au/national/education/all-in-together—197-students-in-one-room-20110605–1fnji.html. Accessed 15 July 2014.

Sydney Morning Herald, 19 April 1934.

Sydney Morning Herald, 11 August 1934.

Tovey, J., 2013. 'Classroom noise linked to poor results', *The Sydney Morning Herald*, 5 December. Available at http://www.smh.com.au/national/education/classroom-noise-linked-to-poor-results-20131204–2yqzo.html. Accessed 15 July 2014.

Wetherill, E. A., 2002. 'Classroom design for good hearing', *The Quiet Zone*, Noise Pollution Clearinghouse, Fall. Available at http://www.nonoise.org/library/qz3/. Accessed 15 July 2014.

Part II
Sound and Violence

6 Throwing Down the Gauntlet

Voice, Power and Sexual Violence in Penal New South Wales

Penny Russell

On 7 May 1832, three men were indicted at the Supreme Court of New South Wales on two counts of rape, specifically for 'violating the person of Catherine Hayes, at the Weather-boarded Hut, on the 27th of January last.'[1] John Spillane and John Costello, both privates in the 39th Regiment, were each indicted as a principal on one count and accessory on the other; Michael Ormsby, seconded from the 39th to the Mounted Police, faced trial as an accessory to both.[2]

The Attorney-General called two witnesses for the prosecution: Hayes herself, and Mary Heylin, who with her husband kept the public house known as the Weather-boarded Hut, a wayside stop for travellers on the road across the Blue Mountains to Bathurst. The evidence of the two women evoked a vivid scene of drinking, idle talk and imprudent badinage that had taken a sudden and violent turn. The three soldiers, Mary Heylin testified, had come in to the house soon after Catherine Hayes arrived, and had bought drinks for her, which she 'took . . . freely.' The soldiers and the woman were in conversation 'in Irish' for the greater part of the night, and the English Heylin could not understand much of what was said. But she thought 'they seemed all very friendly together'; Hayes was 'laughing and talking with them all the time . . . she was merry, skylarking, . . . she was joking and gammoning the soldiers; . . . from her manner and behaviour, one could not think any other than that she gave encouragement to the soldiers.'

The laughter and chat, by Heylin's account, had taken a sinister turn when one of the soldiers laid a wager with his friends that he would 'have the woman that night.' Hayes retorted that she 'would bet them a shilling they would not have her that night.' Not long after this exchange, the three men started to compare their heights by measuring themselves against the door, and one said he would bet half a pint of rum that Hayes was taller than he. Hayes, 'thinking there was no harm,' went over to the door and turned her back so the landlord could measure her against it. In that instant one of the soldiers seized her; they dragged her outside and commenced a violent assault. When the Heylins, hearing her cries, followed them outside, the soldiers drove them back to the house with violent threats. For five hours afterwards, Mary Heylin said, she listened helplessly to Catherine Hayes' cries of 'murder' and vain calls for aid.

In court, Hayes herself gave details of the soldiers' brutal treatment, telling how they had kicked her, beaten her and struck her so hard across the head that she had lost consciousness. For that very reason, she seemed unable to testify positively to the 'violation' of her person. On the vital point of penetration, which was necessary to establish for the capital offence of rape, her evidence was hazy. '[N]othing took place before I became insensible,' she asserted; and 'as long as I had my senses they did not injure me, and after that I don't know what happened.' Mary Heylin testified that she had gone outside and seen one of the soldiers having 'connexion' with Hayes—but added that Hayes had not appeared insensible at the time. She was crying out, and 'must have been aware of what was going on.' Hayes' own insistence that no penetration had occurred while she was conscious thus served to undercut Heylin's testimony on her behalf. Heylin also gave ample evidence of the violent treatment Hayes had suffered, and the bruising visible on her body the next day—but she specifically mentioned bruises only across her back and shoulders. She added that Hayes had been willing to 'make it up with' the men, if they would only give her the money for a new bonnet, to replace the one that had been torn to shreds by their violence.

Trials for the capital offence of rape were a rare occurrence in colonial New South Wales.[3] This one was quickly over. When Heylin's testimony was completed, the Attorney-General rose to address the judge, Chief Justice Sir James Dowling, telling him he could take the case no farther, since the evidence of Hayes, as 'prosecutrix,' had failed to establish the essential fact of penetration. Dowling, having pressed the Attorney-General to state his precise grounds for abandoning a case that had, he said, attracted the public eye, proceeded to address the jury, explaining that the three men must now be acquitted. 'Most undoubtedly, the poor woman has been grievously ill used,' he observed. But since no lesser charges of assault had been made, the only question for the court to consider was 'whether the prisoners are guilty of a *rape*, in the legal sense of the charge.' In the face of Hayes' curious silence on that subject, the case for the prosecution had collapsed.

The failed trial was a result of what Hayes and Heylin said, and what they did not say, to the court. In the aftermath of the trial, there were those who would claim the women had been persuaded—by threats or bribes—to perjure themselves by giving false testimony. Others would assert that the case had initially been—as Dowling CJ put it to the jury—'got up in some quarter or other,' and that the trial had simply exposed the frailty and contradictions in the original allegations.

While political interests and enmities may indeed have helped to bring the case to trial, some influential men in the colony also had an interest in shutting it down. The Police Magistrate at Penrith—to whom Hayes initially took her complaint—was Thomas Edward Wright of the 39th

Regiment, Captain of the company to which the accused soldiers belonged. His discouraging response caused Hayes to let the matter go, and nothing more might have happened had not the incident come to the attention of Sir John Jamison, a Justice of the Peace and resident of Penrith, who had old grounds for enmity towards Wright. Jamison enthusiastically took depositions from staff at the Weather-boarded Hut, and caused a fellow magistrate to take a statement from Hayes herself. On the strength of these depositions, the matter was considered again by a full bench of magistrates at Penrith, who agreed, despite Wright's strong opposition, to refer the case to the Supreme Court. Following the trial, after reading a colourful account of it in the *Monitor*, Jamison wrote a long letter to that newspaper, justifying his own involvement and criticizing Dowling for not having compelled the prosecution to proceed. In defence of his position, he paid the *Monitor* to publish the original depositions, which differ so markedly from the later testimonies of Hayes and Heylin as to lend support to his suggestion that the two women had been persuaded to change their stories before they appeared in the Supreme Court. Jamison's actions may have been, as historian J. M. Bennett has emphatically judged them, 'inappropriate,' and his critical reflections on Dowling 'misconceived.'[4] But in laying open the shifting and contradictory testimony that the case produced over time, he did great service to history. Edward Smith Hall, owner and editor of the *Monitor*, an outspoken critic of government and an enemy of both Dowling and Wright, was happy to lend Jamison his support. He not only published Jamison's material but himself wrote (and published) a letter to Viscount Goderich, Secretary of State for the Colonies, complaining that the case showed only too clearly the tyranny of military authority and the corruption of government in the colony.[5] A blistering exchange between Jamison and Wright was then carried on in the *Monitor*'s columns until early July, each letter revealing more contested details of the affair and its politicization.

The 'true story' of what happened at the Weather-boarded Hut on that fateful evening will never be uncovered. But the numerous, contradictory efforts to see 'justice' done in the case resonate with voices heard and overheard—sometimes listened to with eager attention, sometimes thrust upon unwilling ears, sometimes ignored or rebuffed. The voice of Catherine Hayes herself, in particular, is everywhere: rarely transcribed directly, but instead filtered through the moral and social judgement of auditors who recollected and reported her words verbatim or in summary. Through those filters we 'hear' her laughing, talking, 'gammoning' the soldiers, conversing with them in their shared Irish tongue and laying a bold claim to the inviolability of her body. We hear her screaming, crying, calling for help and begging for mercy; and at last we hear her speaking before judge and jury in the Supreme Court. But we do not know whether the story she told in those intimidating surroundings was her own. Catherine Hayes was certainly willing to speak in her own defence, and others testified to hearing her voice raised in strident self-confidence. But on numerous occasions, also, her voice was stifled, ignored, manipulated, exploited or wilfully misconstrued.

The contradictory testimony surrounding the case is thus crammed with evidence about the sound of human voices, and their impact on their hearers—and so provides the basis for my argument in this essay. The metaphor of voices is a common one in history. As Alan Atkinson has put it, 'hunting for voices is the historian's essential task.'[6] Responding to the burgeoning interest of convict historians in the enumeration, classification and description of convict bodies, he argues that such history may too easily 'collude' with the convict system itself to reduce human beings to mute spectacles. 'Bodies are dispensable,' Atkinson argues. 'Voices are timeless' and 'the only real medium of the soul.' What appears on convict bodies is thus 'contingent on what can be heard in their voices.' But when Atkinson listens for voices, he is listening chiefly for the stories they tell. He has little interest in voices that are 'simply sounds coming from leathern faces and beaten bodies'—the screams and curses elicited by a flogging, for example. The final identity of convicts, he suggests, is to be found 'in what they felt, thought and said over a period of time. Nothing can beat the telling of stories. The unfolding of character, in story-telling, is by far the best proof of humanity.' Thus for Atkinson, the 'voice' that proves humanity is not an embodied, human voice after all, but the metaphorical voice of narration that survives in text and thus achieves a kind of 'serendipitous human contact' between the past and the present.[7]

Atkinson here gives eloquent expression to a common assumption about historians' interest in the voices of the past. The *sounds* of the human voice are still a less common focus of attention, though some important pioneering work has been done by Jane Kamensky[8] for early New England and Joy Damousi[9] for Australia. In this chapter my aim is to build on such work by teasing out, in historically specific contexts, intonations of the human voice and their effects on others—whether building goodwill in jovial conversation with peers, overheard expressing terrible intent, betraying origins in its accent and vocabulary or raised in anger, mirth, pain or fear.

In attending to the sound of the voice, I face the fundamental problem that besets all historians of sound: that, certainly for the colonial period in which I work, we have no way of hearing the sounds themselves, but must work with their representation in text. But as Mark M. Smith has argued, textual sources should not be regarded merely as 'indirect' sources on sound: rather, they are 'direct' sources on how those sounds were heard and the cultural significance that was attached to them.[10] In the case of speech, as Kamensky observes, we work perforce with 'snippets of talk' that have been 'screened at least twice, through the screen of recollection, and through the filter of writing.'[11] But by focusing attention precisely on those filters, acknowledging that what finds its way into text is not the original conversation but a series of impressions about its impact, we may in fact come closer to an appreciation of how people in colonial New South Wales—across differences of gender, power and social position—heard each other, and how they interpreted what they heard. Thus I endeavour to open up what Veit Erlmann describes as a 'twofold process of "hearing another person listen"' in order to 'pick up all these sounds adrift, these echoes, reverberations, hums and murmurs outside or in between the carefully bounded precincts of orderly verbal communication.'[12] We may bridge the gap Atkinson posits between mere *sounds* and narrative *voices*, I suggest, by listening for meanings produced within the 'social relations of talk'[13] and made apparent particularly through mis-hearings, mis-rememberings and ambiguities.[14]

This is quite a different project from attempting to find Catherine Hayes' 'voice,' and proof of her humanity, in the unfolding of her story. In the corrupted, broken, contradictory testimony that found its way on to public record, we can catch only the faintest echoes of the voices of the powerless. While the story of Hayes' rape was by turn repressed and recovered, the voices and agenda of powerful men always drowned out, dismissed, exploited or reinvented the stories that Hayes and Heylin tried to tell. Since both were illiterate, their stories could become text only through the filter of another person's writings. There is no innocent testimony here; we cannot rely on its 'unfolding of character' as a 'proof of humanity'—or, indeed, as proof of anything else. Instead, by focusing on the sounds of voices, we may come closer to understanding what Kamensky has termed 'the politics of speech.' By examining the way people in particular communities used speech, Kamensky argues, 'we begin to uncover something of how they conceived of themselves. We begin, that is, to *hear* their history: to restore the voices, the silences, and the clamor amid which people in that distant world made sense of their lives, day by day.'[15]

In the essay that follows, I explore the human voice in three timbres. First is the politics of speech and community, and especially the way the significance of voices, Irish speech, jokes and wagers shifted within the changing testimony of Hayes and Heylin. Second is the dynamics of violence and power, and the tyranny exercised by the soldiers over an isolated community through the sound of Catherine Hayes' screams. And finally, I explore the fate of Hayes' metaphorical 'voice' as she endeavoured to tell her story to different auditors, sympathetic or hostile. While these explorations illuminate the complex social dynamics of the case, they also tell us something about the politics of hearing and listening in nineteenth-century New South Wales.

'Talking Irish': Sound, Speech and Community

On 27 January 1832, Catherine Hayes, an Irish-born ex-convict, was on her way homeward from Bathurst to Richmond, where she lived with her husband, a convict assigned to service. She began her journey in company with some bullock teams, for protection, but when she stopped for refreshment the teams went on without her. She pursued her way alone and on foot, vainly hoping to come up with her party. Instead, she fell in with a young Irishman, a private in the 39th Regiment, who was going her way. His name—though he never told it to her—was Thomas Brennan, and he was headed for the Weather-boarded Hut, where he was currently stationed with a small detachment of his regiment under the command of Sergeant George Milward. The official duties of the soldiers at this garrison were to escort convict gangs removing from one station to another and to appre-hend 'bushrangers'—a term that designated any convict at large without permission. They were also meant to provide protection to travellers across the Blue Mountains.[16] Whether anyone felt the safer for these soldiers' pres-ence, however, is a matter of doubt.

The Weather-boarded Hut stood on a lonely spot high in the moun-tains, some 50 kilometres by road from Penrith and 100 kilometres from Bathurst.[17] Besides the garrison the only significant building was the 'neat little inn' that stood opposite, at some 180 metres distance: the public house run by the Heylins.[18] In this isolated community, the small military force held significant power—in numbers, authority and sheer weight of arms. It was a power they were only too willing to abuse. Many of the privates of the 39th had been recruited from the peasantry of Ireland before the regi-ment was sent to New South Wales in 1825. For many, soldiering may have offered a last refuge from destitution; a different spin of fortune's wheel might have seen them turn to crime and be transported as convicts to the same place. In status, conditions and freedoms, indeed, their fate was not much better than that of the convicts they guarded. But their duties were relatively light and undemanding, and their military discipline less than perfect. Lonely and bored, and no doubt brutalized by their conditions of service, their chief entertainment seemed to lie in a masculine culture of drinking and gambling. They lorded it, unchecked, over the district, becom-ing not its protectors but its greatest menace.

Of late, rumours had been flying fast. It was said that a convict woman had been 'taken forcibly from the team by which she was travelling and kept in the bush for a whole week.' A free woman, a resident in the area, had been taken and kept for a night. These outrages were freely attributed to the soldiers at the Weather-boarded Hut.[19] Hayes had heard the talk, and knew the dangers of her situation. It must have been with mixed feelings, therefore, that she encountered Brennan, who might as easily have repre-sented a source of danger as protection from it. As it turned out, or so she later deposed, he 'conducted himself very properly.'[20] It emerged from her own and Mary Heylin's depositions that this 'proper' conduct included his offering her money to 'go into the bushes with him'—but when she refused, telling him that she was 'no vagabond, but the mother of ten children,' he did not press the point.[21] They walked on some 9 kilometres together, and arrived safely at the Weather-boarded Hut just after sundown. There, Hayes bought a drink for Brennan and another for herself, ordered dinner and requested a bed for the night. There was no chance now of catching up with the bullock teams before nightfall; she was in any case afraid to go on 'in consequence of the bad name she had heard of the soldiers taking women into the bush.'[22] Her fears were underlined by the entry at about this time of the three soldiers: Costello, Spillane and Ormsby.

It is at this point that significant differences emerge between the way the story was told in the depositions and that contained in the testimony before the Supreme Court. The differences amount to two quite distinct versions of the conversational exchanges, hearings and overhearings that were prelude to a violent rape. The way Mary Heylin described events to the Supreme Court, the violence of the soldiers seemed to arise directly out of their drinking and suggestive talk with Hayes. She told how Hayes accepted several drinks from the soldiers, how she joked and 'gammoned' them—and she emphasized that 'they were talking Irish the greatest part of the night; I did not understand what they said.' It seemed from this that Hayes and the soldiers were conversing either in Gaelic or in so thick a brogue that Heylin could not follow it. Heylin's emphatic repetition of the point—'she was in conversation a long time in Irish'—seemed to indicate her own feelings of alienation or antipathy towards this Irish community, which in turn might explain her apparent lack of sympathy for Hayes.[23] Unable to understand the conversation, she instead drew conclusions about its purport from observation. From Hayes' 'manner and behaviour, one could not think any other than that she gave encouragement to the soldiers,' Heylin said coldly. 'I felt ashamed of having such a woman in my house.'[24]

Heylin here seemed almost to imply that Hayes was fair game for the soldiers: not because she was socially inferior to them, but rather because they were too alike. She and the soldiers formed, together, a disreputable Irish community, bound by shared language and mores. Such cultural distancing, marked by her incomprehension, apparently enabled Heylin to limit her sympathy with Hayes, even while she condemned the conduct of the soldiers. Her testimony veered unevenly between expressions of horror at the violence of the attack and expressions of prim disapproval of the other woman, accompanied by apparent indifference to her sufferings.

Instructing the jury to acquit the three soldiers, Dowling CJ told them that even if the whole case had been submitted to them, he would have directed their attention to certain circumstances that might suggest the men were not guilty of rape. He dwelt with particular energy on Hayes' conduct in 'laying a wager with the prisoners upon her own chastity, before the alleged violence was committed. What could any woman reasonably expect,' he asked, 'when such a Ganntlet [sic] is thrown down to three drunken soldiers?'[25] Thus a jesting wager, overheard by the publican's wife, recollected and relayed by her for the court more than three months later and summarized in an unsympathetic spirit by the chief justice, was represented to the jury as a distinct provocation to rape.

A rather different story was told in Mary Heylin's deposition before Jamison in March—which was backed by statements from Hayes herself,

William Heylin and two members of the staff at the public house: a waiter named Joseph Edmonds and the ostler, John Cook. All the witnesses expressed their sympathy with Hayes, and said nothing to suggest that she had provoked, let alone deserved, the brutality visited upon her.

The depositions hint, indeed, that the soldiers' arrival at the inn was no coincidence—that they came there, with fell intent, precisely because they had heard of Hayes' arrival. As Hayes herself told it, Brennan was just leaving the house when the others arrived, and she heard them outside asking him 'can we have that woman?' When he told them they could not, because he himself had offered her money and been refused, they responded, 'we will have a trial however.' When they entered the house soon afterwards, one of them told Hayes crudely that if *he* had come along the road with her, 'he would have known whether she was a man or a woman, and he would know it before ten o'clock that night.'[26] Hayes' ill-fated wager, in this context, might be read as an attempt to assert her rights without angering men whose express intentions she had already overheard.

Mary Heylin's deposition tells a similar story. In it, the 'Irish' speech seems to be directed against Hayes, rather than including her. 'The soldiers conversed in Irish,' she said, 'when Brennan, who first came in with the woman, said it was of no use, for he himself had offered Hayes money to go into the bush, which she refused.'[27] Both women's accounts thus vividly evoke a sense of the violence that threatened Hayes from the outset, emanating from the soldiers' talk among themselves. Here Heylin's reference to the men 'talking Irish' conveys not an imprudent conviviality between the woman and the soldiers, but rather the men's use of a tongue Hayes understood on purpose to direct their lewd and violent suggestions to her—meanwhile veiling them, albeit imperfectly, from the English staff at the inn. Heylin's deposition also suggested that, far from drinking and joking with the soldiers, Hayes 'became alarmed' from the moment they entered the tavern. Indeed, it was their arrival that prompted her to beg the landlady for permission to remain under her roof that night.

It was not only the two women who heard distinct menace in the soldiers' words. In his deposition, William Heylin told how Spillane had urged him to 'turn the woman out,' which he refused to do.[28] Spillane demanded to know whether she was a free woman; Heylin assured him that she was, and that she had given him her certificate of freedom for safekeeping. He thought the soldiers wished to get hold of the certificate so they could destroy it, 'to make her out a bush-ranger' (and thus perhaps furnish themselves with a pretext to abduct her, in the guise of an arrest). The ostler, John Cook, saw at once that the soldiers were 'bent upon mischief.' There would be trouble, he knew, and he had no desire to get mixed up in it: 'fearful of the consequences, he went to bed.'[29]

Everyone in the inn that evening, the depositions suggest, could hear the threatened violence in the soldiers' voices, in their crude discussions with each other, in their wagers and insulting suggestions. Hayes' voice was raised not in vulgar jest but in anxious entreaty, as she sought to gather around her all the frail protection the inn staff could offer her. If she did converse in Irish with the soldiers, and joke and 'gammon' with them, her apparently light-hearted banter may have masked a rather desperate attempt to assert a sense of community with them, and to remind them of her humanity and right to respect in terms she hoped they might understand. Just as she had told Brennan (untruthfully) that she was a 'mother of ten,' now she reminded the soldiers of their shared Irishness. Later that night she would repeat these appeals to their feelings explicitly—and to no avail. '[T]hey told me to be quiet,' she told the Supreme Court. 'I said "no—if you are all Irishmen, did any of you come from a woman?"' But the soldiers were deaf to such appeals. One 'made answer, that his mother was a cow, and the other replied that his mother was a mare.'[30] Spillane said something Hayes found still more offensive, and that no newspaper would print.[31] 'I cut him in the nose at the time he said that,' she later told the Supreme Court.[32]

Before the Supreme Court, Mary Heylin appeared as a disapproving audience to both Hayes' imprudent conversation *and* the soldiers' horrific violence. On the basis of her testimony, the chief justice could plausibly represent the rape as arising from Hayes' own provocative conduct. In the story told in the depositions, in contrast, Hayes seemed to have entered into a bond of community not with the soldiers, but with the inn staff. The critical differences between the two stories lay in the details, not of what was said, but of *how* it was said, and still more of how it was heard or overheard. What is perhaps most remarkable about these shifting stories, however, is the subtlety of the differences. Mary Heylin, in particular, showed a fine awareness of how social and moral judgements might be made on the basis of speech and listening. By retaining in both stories many details of words spoken, but changing the context in which they were heard, she entirely changed the colour of her evidence about the case.

Sounds of Violence

Just before 10 o'clock—the time Spillane had mentioned in his wager— the three soldiers conferred outside for a moment, then returned and stood near the door, jokingly comparing heights, and thus luring Hayes close to the door and to disaster. Mary Heylin's account of what followed varied little between her deposition and her Supreme Court testimony. The soldiers faced no charges of assault, so her evidence of their violence could not, in itself, harm them. That they had inflicted terrible injuries on Catherine Hayes, indeed, was never in dispute.

It was Costello who seized Hayes at the door, crammed a handkerchief into her mouth to stifle her cries, and dragged her outside. Some 70 metres from the inn, he threw her to the ground and the violent assault began. Hayes resisted in 'every way in her power.' When one of the men—probably

Spillane—'attempted to violate her person,' she fought so hard that she was 'too strong for him'; realizing this, another pinned her brutally to the ground by planting his foot on her face and neck. Still she struggled and tried to bite, temporarily forcing Spillane to desist. The evidence of her resistance could be seen the next day in the 'tufts' that lay scattered on the ground, torn from the shoulders of the soldiers' uniforms, and in the vivid scar across Spillane's nose, where she had slashed him with her nails. But her own body, battered and bruised, and the rent and bloodied ruin of her clothes and bonnet, bore witness to the failure of her struggle. Heylin's deposition was so explicit about the evidence of rape—so Jamison reported—that her words could not be written down: they appear only as a white space in the *Monitor*'s columns. With boots and stick and heavy fists, the soldiers had bludgeoned Hayes into submission.[33]

But not into silence. From the moment they seized her, Hayes' voice was raised against her attackers. As soon as she could free her mouth from the handkerchief that gagged her, she 'called out murder murder, as loud as she could vociferate for assistance.'[34] She pleaded, even argued, with her attackers. And she cried repeatedly for help. Her first cries brought the Heylins and the 'old waiter,' Edmonds, outside. Seeing her on the ground, and one of the soldiers lying on top of her, they 'called out shame.' When Hayes begged William Heylin to help her, however, he replied—according to her own deposition—that 'he was willing to do so, but that life was sweet, and he knew he should be murdered if he interfered.'[35] One of the soldiers then picked up Hayes to carry her further from the house; the other two hustled the Heylins back to the inn, threatening to blow or beat their brains out if they dared to approach again.[36]

One of the striking things about the depositions of the witnesses is their detailed recounting of the words spoken by Hayes, the soldiers and the inn staff. Consisting largely of threats, curses and pleas for aid, they nevertheless invoke the enforced intimacy of proximity in which the violence was enacted. Nothing could more clearly underline the power of the soldiers in this tiny, isolated settlement. Their efforts to remove themselves from within sight and earshot of the inn staff were almost insultingly perfunctory. Their indifference to the proximity of the barracks also suggested that Hayes could expect little help from there.

The Heylins were easily intimidated. The waiter, Joseph Edmonds, went so far as to say loudly that someone should report the crime to Captain Wright. But he paid for his moment of valour with a night of fear. They 'looked up and down for the man that said this,' Mary Heylin later reported, 'and said they would have his head that night or in the morning.'[37] Edmonds was forced to hide in a bed of potatoes. The Heylins and the waiter 'sat up listening to the deplorable cries of the woman in concealed places untill three o'clock in the morning; the soldiers approached the house several times in the night to ascertain if all was quiet.'[38] The ostler, John Cook, who had taken to his bed earlier in the evening, 'distinctly heard them take the woman out of the house, and heard the frightful cries of *murder*.' He listened to her cries for 'about three hours,' he said, but was afraid to move from his bed, 'having heard the soldiers declare, that they would have the life of the waiter.' Mary Heylin said that the woman's cries were so distressing that neither she nor her husband could go to bed 'until she had ceased calling murder.'[39]

'I can scarcely . . . conceive a situation, except it be that of the sufferer herself, more horrible, than to be compelled to listen for five hours to the cries and groans of a woman, exposed to the reiterated violations of four monsters in human shape.'[40] So, after the event, wrote Edward Smith Hall in an open letter to Viscount Goderich, Secretary of State for the Colonies. Hall attributed the passivity of the inn staff to the fact that the military were 'the privileged class of this colony' and justifiably 'feared in any public irregularity they may commit.' Knowing that military officers exercised civil authority as magistrates, the inn staff would have 'dreaded the danger of interfering with the military *at all*, in any way, or by any manner of means' and feared to set their future livelihoods, as well as their present safety, at risk. Moreover, he added patronizingly, to 'interfere with men under such circumstances, would require a degree of courage, not to be found in the class of persons who saw and heard the catastrophe I am describing.' But dead though they might be 'to that moral courage, which in such circumstances would induce some to risk life, they were still human beings, and could not but sympathise with the wretched sufferer.'

Hall's assumption of a universal human response to the sound of human suffering may need some qualification. Cook and Edmonds were ticket-of-leave men; the Heylins (though they did not advertise the fact) were also ex-convicts. This may not have been their first experience of helplessly witnessing the pain of others. Although officially floggings were not administered publicly after 1820, 'they were nevertheless carried out as exemplary exercises before captive convict audiences.'[41] Convicts may never have learned to close their ears to the sounds of human suffering, but they must have developed strategies to endure it. The rape of Catherine Hayes was different, in that pain was inflicted not as a matter of state-sanctioned discipline, but in ways that were manifestly illegitimate, though equally incontestable. A feeling that they should, though they dared not, intervene may have exacerbated the suffering of the auditors. Thus a handful of drunken soldiers exercised their unchecked and brutal sway over those within earshot, in an auditory circle defined by Hayes' unanswered screams for help.

That they did suffer seems evident from their subjective estimates of how long the cries continued. To Cook it felt like three hours, to Mary Heylin five. Over in the military barracks, however, allegedly nothing was heard until a convict grass-cutter alerted one of the soldiers, telling him 'that their men had a woman in the bush.'[42] According to his own deposition, Stephenson had been aware of nothing untoward until Thompson spoke to him.

But when he went outside 'to the end of the Barrack' he, too, could hear the woman's cries. He took off his shoes and approached with caution, then watched and listened, according to his deposition, while Thomas Brennan (whom he identified by his voice) violated the woman 'three or four times' and 'beat her most unmercifully with a stick.'[43] Only when Brennan was 'done with the woman' did Stephenson intervene and persuade her to return to the barracks with him, where he wrapped her in blankets and guarded her with his gun for what remained of the night. By his account, the barracks were closer than the inn to the spot where Hayes lay, making the failure of any other soldiers to hear Hayes' screaming all the more remarkable. Captain Wright would later claim that this was because the whole incident was much briefer than Heylin and Cook had made out: Stephenson, he said, had insisted that it was not later than 10 or 11 o'clock when he went to the woman's assistance. If the cries sounded relentlessly and interminably in the ears of the inn staff, the soldiers in the barracks perhaps had good reason to wish to minimize, in recollection at least, both their duration and their volume.

Distorted Echoes: Hearing Catherine Hayes

At the Supreme Court trial, Mary Heylin had stated that Hayes had agreed to 'make it up' with the soldiers for the price of a new bonnet. Dowling CJ told the jury, damningly, that this showed Hayes' 'own sense of the injury.' Coupled with the 'alleged staleness of the charge,' it all suggested that Hayes herself had not been greatly interested in pursuing the matter at the outset.

Again the depositions tell a different story. Early the next morning Catherine Hayes complained to Sergeant Milward of the treatment she had received. Since the only man she could identify by name was John Spillane, Milward duly sent him in custody to Emu Plains, where his fate would be determined by his superior officer, Captain Wright. Mary Heylin, meanwhile, tended to Hayes' hurts, and agreed to go to Penrith to support her in her complaint against the soldiers. Far from 'looking over the matter' for the price of her bonnet, Hayes seems to have been determined, at this stage, to see justice done. She was sufficiently recovered to make her own way to Penrith next day. There she spoke to the chief constable, but was told that she would have to wait for the next court day to lay her complaint before the magistrate. She went home to Richmond, returning just over a week later to tell her story to Wright. In the meantime, the Heylins had come to Penrith, made their own statements before Wright and returned to the Weather-boarded Hut. When Hayes appeared before Wright and his fellow magistrate George Druitt, therefore, she was alone and unsupported.

There are several accounts of what transpired next—all made weeks after the event and at the instigation of various concerned parties, and none of them, therefore, particularly reliable. According to Druitt, Hayes' 'manner was exceedingly insolent and contemptuous towards the Bench; she frowned upon Capt. Wright, and behaved in such a manner, as induced me to think she had been drinking.' Although Wright signified his willingness to give her justice, if she would point out the men who had injured her, Hayes had responded in a manner 'highly provoking and insolent, concluding with these words, and putting her arms a-kimbo—"I'll go to your betters, who will see me righted." '[44]

Alex Fraser, Clerk to the Bench, told the story differently. He said that Hayes had appeared 'eager and anxious to have her case investigated,' but when Wright told her that he would send for the offenders if she could give him their names, but 'could not think of bringing down the whole detachment,' she had become violent in her manner. She 'could get no justice there,' she declared, and left the court, 'evidently chagrined and disappointed.'[45] Constable William Pickett gave a still more colourful account of the exchange, saying that after Hayes made her complaint to Captain Wright, 'he told her, he did not believe one word of it, or any part of her statement; the said Mrs. Hayes then replied, I will then make reference to a better Judge, on which Captain Wright told her she might go to the devil if she liked.'[46] Hayes' own later deposition was transcribed in relatively neutral terms: she stated 'that she went to Penrith Court-house for the purpose of complaining, but in consequence of deponent having no money, and being advised not to prosecute the parties, she returned home, thinking it better to say nothing about it.'[47]

Whether Hayes' drunken insolence was thwarted by Wright's voice of reason, or her legitimate complaint stifled by his arrogant determination to protect his company from scandal, the central point is clear—that her brief encounter with the magistrate's court convinced her that, after all, silence was her most prudent course. There, indeed, the matter might have rested had not Sir John Jamison travelled in March across the Blue Mountains and stayed for a night at the Weather-boarded Hut, heard the story from the Heylins and taken down the depositions that he later published. Jamison was an active, encouraging listener, keen to elicit precisely the stories Wright had not wished to hear. We will never know for sure whether this revival of the story was prompted—as Jamison himself suggested—by Mary Heylin's increased fears for her own safety, now the soldiers had learned that they could rape with impunity, or whether—as Captain Wright would fiercely imply—it was deliberately stirred up by Jamison himself, in pursuit of a longstanding vendetta against Wright.[48] Either way, it is clear that Catherine Hayes' story was never elicited from her or other witnesses by a neutral court, keen to learn the truth while events were fresh. Instead, it was by turn suppressed and uncovered, suppressed, then uncovered again, by men who had not witnessed the rape, but who nevertheless took responsibility for turning it from an event into a discourse. Only because of vested masculine interests in debates over honour, integrity, justice and power did the pitiful story of Catherine Hayes' suffering find its way first into depositions and then into the newspapers, and thus into historical record. Because of them, her voice, and the story she tried to tell, echoes faintly through words and media that were never her own. Such appeals for help as she made to the law proved as futile as her screams for aid during the assault. Too many men had their own story to tell, their own interests to serve. Her voice and her story, like her battered body, were pressed into the service of others.

Notwithstanding the tantalizing glimpses of community interaction and everyday speech it offers us, the 'proof of humanity,' in Hayes' case, does not lie in the 'unfolding of character' through story—mainly because the story was never her own. The stubborn silences and inherent contradictions in these accounts cannot be evaded or overcome, nor can we grasp with confidence the meanings that any one player may have attached to the events of that night. Empathy is not possible here, nor do these broken stories offer a basis for the 'serendipitous human contact' that Atkinson desires. Hayes' narrative voice was too effectually silenced. What we do find is evidence of the meanings attached to conversational style, evidence of the power of those who listened and those who refused to listen and evidence of the use of deliberate ambiguity and mishearing to re-present the rights and wrongs of a matter. But we also find, present in every account, the 'deplorable,' 'horrible,' 'frightful' and 'distressing' sound of Catherine Hayes' screaming. The textual traces of those cries do not override the complexity of our engagement with the past, but they are the most powerful reminder in this case of its irreducible humanity. The sounds emanating from Hayes' 'beaten body' can still catch at the helpless witnesses of her suffering, torturing them with awareness of her humanity and the resistless power of her attackers.

Notes

1 *Sydney Gazette*, 2 June 1832, p. 3.
2 The *Sydney Gazette*'s account of the trial (*R v Spillane, Costello and Ormsby*, 7 May 1832) may also be found online at: http://www.austlii.edu.au/au/cases/nsw//NSWSupC/1832/32.html. An earlier and colourful account had been published in the *Monitor* (12 May 1832); the *Sydney Herald* (14 May 1832) reported only the indictments, and that 'the case broke down'; the *Australian* (18 May 1832) treated it with greater brevity still. The *Monitor's* account prompted Dowling to have his own record of the Supreme Court trial published in the *Sydney Gazette* (2 June 1832) and sparked a longstanding press debate about the rights and wrongs of the case. Where not otherwise specified, quotations in this chapter are from the *Gazette*.
3 On sexual assaults and the politics of punishment see Bongiorno (2012, especially Chapter 1) and Kaladelfos (2012).
4 Bennett 2001, pp. 79–81.
5 Hall 1832.
6 Atkinson 1999, p. 25.
7 ibid., pp. 26, 27.
8 Kamensky 1997.
9 Damousi 2010.
10 Smith 2004, p. xv.

11 Kamensky 1997, p. 10.
12 Erlmann 2004, p. 18.
13 Kamensky 1997, p. 10.
14 See Carter 2004, pp. 43–63.
15 Kamensky 1997, p. 10.
16 Hall 1832; Bennett 1834, p. 102.
17 The site of the present-day town of Wentworth Falls.
18 Hall 1832; Bennett 1834, p. 102.
19 *Monitor*, 2 June 1832, p. 2.
20 *Monitor*, 30 May 1832, p. 3.
21 Ibid.
22 Ibid.
23 O'Farrell (1987, pp. 26–27) argues that Gaelic was considered, in Australia as in Ireland by this time, a 'badge of poverty and failure' and a 'bar to the world of power.' Though many Irish arrived almost totally ignorant of English, they abandoned Gaelic as soon as they could.
24 *Sydney Gazette*, 2 June 1832.
25 Ibid., p. 3.
26 *Monitor*, 30 May 1832, p. 3.
27 Ibid.
28 Ibid.
29 Ibid.
30 Ibid.
31 Indeed, the *Sydney Gazette* would not print any of the soldiers' comments, but left blank spaces in their transcript of the testimony.
32 *Sydney Gazette*, 2 June 1832, p. 3.
33 *Monitor*, 30 May 1832, p. 3.
34 Ibid.
35 Ibid.
36 *Sydney Gazette*, 2 June 1832, p. 3.
37 Ibid.
38 *Monitor*, 30 May 1832, p. 3.
39 Ibid.
40 Hall 1832.
41 Evans and Thorpe 1998, p. 25.
42 *Monitor*, 30 May 1832, p. 3.
43 Ibid. (Hayes herself did not name Brennan as one of her assailants. By the time Stephenson made his statement, Brennan was facing court martial for firing at Sergeant Milward in a drunken rage. He was executed on 6 April [*Australian*, 6 April 1832, p. 3; *Monitor*, 9 June 1832, p. 4]).
44 *Monitor*, 9 June 1832, p. 4.
45 Ibid.
46 *Monitor*, 2 June 1832, p. 2.
47 Ibid.
48 *Monitor*, 30 May, 2 June and 9 June 1832.

References

Atkinson, A., 1999. 'Writing about convicts: Our escape from the one big gaol', *Tasmanian Historical Studies*, 6(2), pp. 17–28.
Australian [Sydney], 6 April and 18 May 1832.
Bennett, G., 1834. *Wanderings in New South Wales, Batavia, Pedir Coast, Singapore and China; being the journal of a naturalist in those countries, during 1832, 1833, and 1834*, Vol. 1, London: Richard Bentley.

Bennett, J.M., 2001. *Sir James Dowling: Second Chief Justice of New South Wales 1837–1844*, Sydney: Federation Press.

Bongiorno, F., 2012. *The sex lives of Australians: A history*, Melbourne: Black Inc.

Carter, P., 2004. 'Ambiguous traces, mishearing, and auditory space', in V. Erlmann, ed., *Hearing cultures: Essays on sound, listening and modernity*, Oxford: Berg, pp. 43–63.

Damousi, J., 2010. *Colonial voices: A cultural history of English in Australia, 1840–1940*, New York: Cambridge University Press.

Erlmann, V., 2004. 'But what of the ethnographic ear? Anthropology, sound, and the senses', in V. Erlmann, ed., *Hearing cultures: Essays on sound, listening and modernity*, Oxford: Berg, pp. 1–20.

Evans, R. and Thorpe, B., 1998. 'Commanding men: Masculinities and the convict system', *Journal of Australian Studies*, 22(56), pp. 17–34.

Hall, E.S., 1832. 'To the Right Honourable Lord Viscount Goderich, H.M. Principal Secretary of State for the Colonies &c. &c.', *Monitor*, 30 May, pp. 2–3.

Kaladelfos, A., 2012. 'The politics of punishment: Rape and the death penalty in colonial Australia', *History Australia*, 9(1), pp. 155–175.

Kamensky, J., 1997. *Governing the tongue: The politics of speech in early New England*, New York and Oxford: Oxford University Press.

Monitor [Sydney], 12 and 30 May, 2 and 9 June 1832.

O'Farrell, P., 1987. *The Irish in Australia*, Sydney: New South Wales University Press.

R v Spillane, Costello and Ormsby, 7 May 1832. Available at http://www.austlii.edu.au/au/cases/nsw//NSWSupC/1832/32.html. Accessed 7 September 2014.

Smith, M.M., 2004. 'Introduction: Onward to audible pasts', in M.M. Smith, ed., *Hearing history: A reader*, Athens, GA: University of Georgia Press, pp. ix–xxii.

Sydney Gazette, 2 June 1832.

Sydney Herald, 14 May 1832.

7 Startling Reports

Gunfire as Social Soundscape in Early Colonial Australia

Diane Collins

This chapter listens to gunfire as an element in the social soundscape of early colonial Australia. As acoustics, the distinctive reports of colonial firearms occurred in many contexts. First and most obviously, as the sounds of Indigenous massacre and dispossession. Within the European population, the strongest associations were also with murder and mayhem, since gunfire announced, so often, the occurrence of crime, distress and death. While such sounds were fundamental to the fact and experience of gunfire in colonial Australia, they are not the subject of this chapter. Instead, the focus is on how the sound of gunfire was used to express community solidarity and, at the same time, to contest the boundaries of civility. Here, the term 'civility' is used to mean less courtesy than larger representations of desirable social codes and behaviours. I first became interested in the many social uses of gunfire when researching the sounds of the Australian goldfields. This analysis broadens that earlier inquiry by listening to the longer context, to the ways in which, from the beginnings of European settlement, gunfire was manipulated, as both presence and absence, to modify and impose identities and right conduct.

Gunfire as Celebratory Sound

Colonial Australians came from a European culture into which was deeply etched a tradition of affirmative or celebratory gunfire. Gunfire announced military victories, royal births, deaths and marriages and any number of great occasions. When Waterloo Bridge opened in London in June 1817, for example, arrayed on the bridge's flagstones were a large cannon and artillery pieces firing a grand salute of 202 guns, the number captured from the French in the famous battle commemorated by the bridge.[1] The use of gunfire at such moments was, of course, an expression of aesthetic as much as political power. Cannon fire was then among the loudest human-made sounds, so the very magnitude of the sound functioned to announce the dominance of a nation, or of a particular social order, in this case the British state.

From the beginnings of settlement, celebratory gunfire was a crucial part of cultural transfers to Australia, especially in terms of recreating an etiquette of deference. The anniversaries of royal births, accessions and coronations were each denoted as 'fixed' days for firing gun salutes. In 1799, for example, the birthday of King George III was proclaimed in the still infant colony by a military parade and then the sounds of three volley fires from the New South Wales Corps followed by a royal salute of 21 guns from batteries of soldiers and sailors stationed in Sydney Cove.[2] George III may have been widely credited with the loss of empire, with 'repeated injuries and usurpations'—to quote from the *Declaration of Independence*—but in New South Wales these guns sounded expanding British power and the growing capacity of monarchy to represent the morality of colony as well as nation.[3] In a penal settlement, under naval governorship, such gunfire also usefully displayed the unity of the military and civilian populations.

In its function as civic announcement, official gunfire extended far beyond the annual event cycle of royal occasions. The routine arrival and departure of boats carrying governors and other dignitaries were also marked sonically. On first landing and on departing, the governor was saluted with 17 guns, the lieutenant-governor with 15. Their visits and departures from ships of war received the same acknowledgement. Beyond the governors, ambassadors, envoys, consuls general and consuls were among the offices also entitled to a prescribed number of firings.[4] Although this acoustic theatre was structured by a predetermined protocol, it impressed on citizens' as much as sailors' and soldiers' long-standing notions of order, respect, rank and deference.

Across time, the annual round of public gunfire was enlarged. Its sounds began to connote a retrospective celebration of a new community's shared history as well as old hierarchies. Each 26th of January, the moment of 'first landing' was celebrated sonically. Ships in the harbour fired their tributes, while from the Dawes Point battery, a number of minute guns equivalent to the years of settlement fired at midday. This sound, one commentator observed, 'involuntarily compels the contemplative mind to take a retrospection of years now beshrouded in temporary forgetfulness.'[5] When the Port Phillip district separated from New South Wales to form the state of Victoria in 1850, the residents likewise expressed their joy by illuminating every house and discharging fireworks, while 'from all quarters the reports of firearms resounded for some hours.' Since good order was maintained throughout these celebrations, such civic gestures spoke in 'highest terms' of the citizens' capacity for right 'conduct.'[6] Across time, the firing of 'loose and random salutes' became 'the favourite method of doing honour to any festive occasion,' Christmas and New Year, most notably.[7]

In these expressions, the sound of gunfire had ambiguous connotations. On the one hand, rather than being heard as death, destruction and private calamity, these sounds were fundamental to the values meant to underlie social and personal relations in the new colonies of Australia. Cannon or musket fire was not noise in the sense of unwanted sounds but a basic acoustic of belonging. As a ritual sound, celebratory gunfire also contributed that coveted sense of an ordered world while intermittently providing useful relief from the everyday expectation of restraint. On the other hand, the civil uses of gunfire within colonial society raised questions about the boundaries between acceptable and unacceptable sound, and therefore about the boundaries of civility and social order. Gunfire did not always sound affirmative messages. In the first half of the nineteenth century we can hear a growing sensitivity—not to gunfire on the frontier or in remote country towns but to the sounds of gunfire in Sydney and its nearby settlements, to the sound of gunfire on the public roads and streets.

Gunfire and the Limits to Acceptable Acoustic Behaviour

While citizens' random firing of guns for pleasure was ubiquitous by the 1850s, this habit long preceded the gold discoveries of that decade. As early as 1810 the acoustic boundaries of acceptable sound were being deeply contested, as standards of civility were asserted by authorities and the emerging respectable elite. In that year, for example, the government issued a public notice on the 'use of firearms.' Attention was drawn to the practice, 'daily increasing,' of persons not only carrying guns but shooting in the immediate neighbourhood of Sydney town. The practice was said to be objectionable on two grounds: as a 'risqué' to individuals going about their private concerns and as a 'violation of decency.' In future, all persons so offending were to be deprived of their guns and all 'peace officers' were to assist magistrates in this matter. Four months later, the notice was amended to draw special attention to the 'firing or shooting of guns on Sunday' and, in September 1811, the order was extended to the fringe settlements of Windsor and Parramatta.[8]

While such regulations aimed at curtailing the physical danger inherent in random gunfire, a larger concern was silencing a sound so readily heard as civic disorder, as an absence of regard for others and as a failure of that restraint to which all notions of civility were bound. It was also soon clear that these regulations fell far short of their intended effect. In 1822, for example, the *Sydney Gazette* published an extended commentary on the flouting of the prohibition against discharging firearms 'in the town of Sydney, or within one mile thereof.' Magistrates complained, the writer stated, of the 'repeated infraction of this order' and of the failure to bring to justice these 'violators of the public peace.' Yet, there was, the commentary continued, no regulation that required more implicit obedience. Significantly, while the paper was prepared to tolerate the infringement of the regulation during daylight hours, it drew a line at the firing of 'well-loaded' pieces 'not only after the evening has far advanced, but at any hour of the night.' This was repeatedly and systematically the case in lower George Street, much to the 'annoyance of the inmates of every dwelling in the neighbourhood.' Nor could there be much doubt as to the character of the perpetrators. Daylight firings were produced by 'impertinent' and perhaps 'idle' fellows, but night discharges were the work of poltroons seeking self-amusement and gratification from 'drunken associates.'[9] Put in this way, gunfire was more than an urban nuisance, it was a noise that sounded idleness, unrestrained alcohol consumption and male immaturity. A few years later, the sound of young men amusing themselves by firing guns in this district was explicitly set against 'the terror of the female inhabitants of the neighbourhood.'[10]

The connection between the sound of random gunfire and dissolute manhood continued and extended far beyond lower George Street and the settlement of Sydney. Within four years of the establishment of the Swan River Colony in Western Australia, the local magistracy were cautioning citizens against the 'extremely dangerous and illegal' practice of discharging firearms in public places and at most unseasonable hours. Such actions amounted to both a physical and acoustic nuisance since there was 'the attendant alarm, and breach of the Public Peace and Quiet.' Evidence of this had occurred in June 1832, when the sudden discharge of firearms by 'drunken idlers' celebrating the anniversary of Waterloo spread panic throughout the town. The natural effect of such false alarms was also to 'destroy all mutual confidence between the residents' while, from the 'incessant reports of firearms, to which we have been too long accustomed, the timid and excited are kept in a constant state of needless uneasiness, while the generality of the inhabitants contract such an habitual disregard of that most effectual mode of alarm, as may be productive of very serious results.'[11] From Hobart, Launceston and Adelaide came the same complaints with the 'annoyance' (rather than danger) of 'letting-off fire-arms in the town.'[12]

Censure of gunfire in these contexts is hardly surprising since it infringed social compacts in relation to the night as a time of rest and the right to claim quiet as a fundamental component of civilized social relations. But gunfire also became the sound of a much more controversial grievance. This is not to say that, as an affirmative expression of public behaviour, gunfire was always heard as appropriate or uncontentious. On occasion, the sound of gunfire became a means to contest notions of acoustic propriety, the behaviours allowable in new societies and to free men and women.

Gunfire and the Godly

About 9 kilometres southwest of Sydney Harbour is a watercourse named after its discoverer: Cooks River. The stream is 23 kilometres long. In 1840, the banks were marshy and well timbered. On any Sunday, a congregation worshipping at the district church heard, outside, the sounds of gunfire repeatedly ringing out. The rest of this paper is about this acoustic encounter, its meaning and its consequences.

First, one consequence. Late in 1840, Governor George Gipps tabled a petition from some 35 'respectable gentlemen'[13] of the Cooks River district of Sydney, concerned that their locality was 'almost daily . . . disturbed by the sound of fire arms, discharged . . . for the purpose of enjoying the amusement of shooting.' People travelled to the vicinity to shoot pigeons, parrots and other birds. On Sundays, in particular, not only did such sounds frighten the horses carrying the petitioners to church but, more objectionably, the hours of divine service were 'disturbed and interrupted by the constant and unseemly noises arising from the practice.' To counter this sonic profanity, the petitioners wanted a legislating of civic restraint: a prohibition on discharging firearms on the Sabbath and 'on any day' within a specified distance of a public road or residence.[14] No reference to a particular religious establishment was made, but the complaint must have come from attendees at St Peters Anglican Church in the then newly created parish of Cooks River. At this time, St Peters was the district's lone house of worship, only consecrated the previous year.

Sundays had long raised issues of sound and silence. To some Sabbath churchgoers, passing carriage wheels could give offence while, later in the nineteenth century, there were churches in the United States strenuously objecting to the auditory intrusion of nearby railway lines. But at the Cooks River, a particularly acute acoustic collision was taking place. On so many grounds, this gunfire could be heard as disruption. In structure, the sound was loud, repeated and random. While its origins were secular, unlike cannon fire, Cooks River gunfire could easily be dissociated from the community since it voiced no identifiable purpose beyond private benefit. And the sound was entirely incidental to intent.

The specific listening context then further amplified this gunfire's power to disrupt expectations of appropriate acoustic etiquettes. First, the particular structure of worship: since no prayer books were then in use at the church, the minister effectively 'read' the service to the congregation. So the primacy of the highly ordered and softer sounds of speech and prayer, words listened to, spoken or thought as inner contemplation made the gunfire's frequent, fitful intrusions all the louder. There was also little countervailing sound in the way of sacred song or instrumental music. The church did have a seraphine, an early reed organ, and some hymn singing took place. But music was then an entirely subsidiary activity within Sydney's Anglican churches. While sermons were often reported in detail, church music received almost no attention.[15]

This is not to say that the church had no louder acoustic ammunition. A bell from the Whitechapel Foundry that cast Big Ben and the Liberty Bell hung in the steeple. The bell should have guaranteed St Peters auditory domination of the surrounding community—in the same way as the elevated situation of the church enabled it visually to command the region.[16]

But the hunters' gunfire essentially neutralized the sound of this bell. This effectively decentred the church as the acoustic focus of the community and, in so doing, gave authority to independent civil behaviours.

Beyond the immediate auditory context of the church, the larger sound-scape of the Cooks River also encouraged the petitioners to hear public gunfire as a transgressive social sound. In this, topography played some part. It does not seem entirely accidental that the area was a river valley. As a natural amphitheatre, the valley was an acoustically vibrant space, a space in which sound echoed. And because rivers—like waterfronts—often consti-tute the edge of a community, this also encouraged hearing contestation. At least one minister of St Peters in this period clearly saw the river as a bound-ary between the church and what he described as the 'wild and godless' place to the southwest, the area between the Cooks and Georges rivers.[17]

For both the new church and its congregation, the control of public sounds was a means of controlling public space. It is especially clear that, in excluding certain types of sounds, the petitioners sought to exclude cer-tain types of people. The areas around St Peters church were sites of new settlement, so it is difficult not to hear the complaint as signalling social conflict over the nature of Sydney's emerging suburban regions. Though then still sparsely settled, the Cooks River district had a preponderance of well-to-do landholders with large estates on which were built substantial houses. Many of these gentry attended or supported St Peters. As testimony, so many of the colony's 'principal aristocracy' attended the church's con-secration that—except at the Governor's Levee—the 'show of carriages' at the event was thought to have 'never been exceeded' in the colony.[18] In con-trast, the petition denoted shooters as 'the lower orders of the community.'[19] While the church made control of gunfire a means of silencing public space and public behaviours deemed rude and socially disruptive, its parishioners were just as clearly seeking a sonic privatization of the river's banks and open areas. This was an especially contentious act as its target was poorer people who sought access to the public amenity of the district on their only available day of leisure, the Sabbath.

But how much and what kind of resonance did this gunfire have beyond the region of Cooks River? In particular, how was this gunfire heard by the colony's governing elite? Legal advice, perhaps dubiously, informed the Governor that no existing colonial or English law specifically prohibited *shooting* on the Sabbath. This meant there had to be majority Council sup-port to pass specific legislation.

Unsurprisingly, the petitioners' most ardent proponent on the council was the Anglican Bishop William Broughton, who personally testified that on Sundays 'almost every third person' he met was 'armed with a musket or fowling piece.'[20] To Broughton, gunfire inevitably sounded the limits to, if not the failure of, the Anglican voice in colonial New South Wales. Espe-cially as, only four years earlier, the Church Act of 1836 effectively dises-tablished the Church of England in Australia. With the free population also

increasing, as it was from the 1830s, opportunities to assert clerical authority were seized upon and demands surrounding the keeping of the Sabbath
already had a long history in the colony. As Sunday desecration, artillery
fire had previously provoked Broughton's ire. When a ship in Sydney Harbour put on a display of booming cannon on a Sabbath in 1831, Broughton
angrily demanded public prosecution to avenge this 'insult to Government
and Religion.'[21]

But for most Council members, the sound of shooting was acoustically and politically juxtaposed not against an idealized Sunday quietude
but against the divisive noise of sectarianism. While the Attorney-General,
for example, had no direct knowledge of the Cooks River region, he noted
that, outside of Sydney, in the Illawarra district (a region he knew well), the
inhabitants made 'no distinction whatsoever between a Sunday and a Monday' and happily shot at virtually anything that flew. Perhaps mindful of the
popularity of such leisure patterns, the Attorney-General believed that the
gradual influence of religion rather than force was the best way of ensuring
the 'proper' observation of the Sabbath.[22] After listening to the discussions,
the editor of one Sydney newspaper expressed relief that the Council had
escaped a dose of the 'Sir-Andrew-Agnew-mania,' a reference to the Seventh
Baronet of Agnew and his four failed attempts, across the 1830s, to persuade the House of Commons to pass Sabbath Observance bills. The paper
supported most members of the executive in advocating cultural change
rather than the implementation of a legal regime that invited oppression in
the hands of 'fanatical' administrators.[23]

Although colonial governments had readily legislated against the indis-
criminate firing of guns in towns, the political elite proved notably nervous about risking the civil uproar that might occur if legal restraints were
imposed on the sound of gunfire that was outside of town limits and a
product not of personal disorder but of the purposeful sport of shooting
birds. In these terms, most Council members also heard the proposal as an
attack on the sounds of the 'poor' amusing themselves.[24] Nevertheless, on
Bishop Broughton's initiative, the Sunday Shooting Bill went to committee. The subsequent report not only supported the proposed legislation but
recommended restrictions on Sunday travel and labour as well as 'all other
riotous pastimes,' a category that included cock and dog fighting, breaking-
in horses and gambling.[25]

By then, an aural commotion extended well beyond the Legislative Council. Sabbatarians across the colony began to hear religious opportunity as
much as impiety and indecency in Sunday gunfire and its suppression. At
Sydney's Scots Church, five Presbyterian ministers across five consecutive
Sundays each delivered a thundering lecture on Sabbath observance.[26] Some
396 persons, magistrates, clergymen and other inhabitants of Melbourne
sent a petition supporting the expanded bill.[27] In Geelong, one newspaper
letter writer pointedly complained less of the number of sportsmen shooting on Sunday than the fact that these Sunday sportsmen lacked acoustic

manners, the 'decency' to remove from within the 'hearing' range of those attending a place of public worship.[28]

Within the Council, the move to turn a bill against a specific aural 'nuisance' into a wide- ranging Sabbath Observance Bill met with no enthusiasm whatever. The Governor attacked the enlarged bill on the basis that, in language and manners, Sunday was a day of festival, albeit a holy one. The proposed bill thus made 'the pursuit of happiness itself an offence.'[29] This extended arguments already made against Sunday shooting. The idea that the 'religious man ought not be interrupted in his devotion' was heard, by some, as an attack on 'every holiday apprentice, every bird-stuffing barber, and every amateur of Ornithology'[30] and thus on 'innocent amusement,' which, in itself, constituted a 'proper observance of the Sabbath.' Acknowledging certain defeat, Bishop Broughton withdrew his amendments. As a compromise, legislation to prohibit only 'shooting for sport or pleasure on Sunday' was consented to in September 1841. Sunday hunting was still legal. This was the first specifically Sabbatarian law passed in colonial Australia, and it became part of the legal framework of all the subsequently formed eastern states. Gipps reported back to Britain that he gave no encouragement to efforts to convert the legislation into a 'Sunday Observance Bill.' He had simply sought to stop a practice that had 'become very general.'[31] For Gipps, acoustic quiet, order and civility were one. Nevertheless, though there was little desire for colonial Sundays to replicate Scotland's day of 'gloom,' the sounds of lower class enjoyment were silenced in favour of sectional interests seeking to remake the auralities of colonial popular culture by recasting widely accepted public sounds and behaviours less as Sabbath desecration than private nuisance. In these terms, gunfire represented less disorder than a breaching of the acceptable limits to the individual assertion of auditory liberty.

Gunfire and the Social Soundscape of an Emerging Democratic State

In practice, however, enforcement was less than rigorous. Within a few years, the British and colonial press were reporting that the law to end Sunday as the 'tradesmen's day of shooting' was widely 'evaded.'[32] It was 1845 before the first sportsman was charged under the Act in the country district of Goulburn, and for the breach the magistrate sought the lowest penalty. The *Sydney Morning Herald* reprinted the first two clauses of the Act lest other members of the public be taken unawares.[33] In 1847, in Geelong, the press observed that it had been so long since a conviction for Sunday shooting had taken place that the constabulary 'must have forgotten the existence of such an Act.'[34] Especially in country areas, police turned the other ear.

Whatever the level of enforcement, the Act passed at a time of evident social concern with public gunfire. In 1854, duelling was banned within the colonies. In 1845, in Melbourne, a parade by members of the Orange

Confederation was heard less as the public espousal of a political cause than a moment of significant civic disorder. A central constituent in this disorder was not just the presence of men 'armed with guns and pistols' but the fact that they 'kept up an almost continual booming of firearms until morning.'[35] The incident led to the passage, in October 1846, of the Party Processions Bill, which outlawed men bearing, wearing or having firearms as well as carrying banners and emblems likely to provoke animosity. Although the contexts were very different, both the Sunday Shooting and the Party Processions bills acknowledged fears that civility's essential quiet was threatened by gunfire as a sectarian sound.

The Acts are also a measure of the growing circulation of guns within the colonies. Before the 1830s, few civilians are thought to have had access to firearms, for reasons both of cost and the colonial governors' fears of convict revolt.[36] Increased circulation was encouraged by the ending of transportation to the east coast in 1840, the trend towards cheaper firearms and the gold discoveries of the 1850s. All gold seekers were advised to carry guns, and the rushes consequently saw the large-scale import of revolvers, American handguns especially.[37]

With greater availability of guns and a relaxation of the social order in such a rapidly expanding society, gunfire became one of the most emblematic, though forgotten, sounds of gold rush Australia.[38] Visitors to the goldfields found the noise of gunfire every evening so ubiquitous that many drew comparisons between the diggings and a war zone. While the causes of this 'abominable din' were various, it meant that, as an acoustic gesture, gunfire not only moved from the margins to the centre of communal life but helped constitute the goldfields as an 'experimental democracy of sound.'[39] In this experiment, gunfire meshed with the prevalence of swearing, blunt greetings and direct, unapologetic conversation to voice a freer, more autonomous citizenry defined, in part, by a rough but 'spontaneous civility.' It was this 'spontaneous civility' which was also observed among the more distinctive features of this period.[40]

While gold undoubtedly contributed to a loosening of the formal expectations attached to social relations on the diggings, acoustic exuberance was never limited to the goldfields. When Mrs. Charles Clacy arrived in Melbourne in 1852, her first night in the city was made memorable by certain 'unmentionable' sounds coming, apparently from a nearby room, the incessant barking of dogs and 'revolvers . . . cracking in all directions until daybreak.'[41] If reports in the *Argus* are a guide, in these years Melbourne's courts saw a succession of offenders charged with discharging firearms in the city's streets.[42]

The post-gold period was widely heard as quieter.[43] Arguably, in the second half of the nineteenth century, two developments helped contain the potential for gunfire to sustain a more aggressive rights culture in which the sound was heard as the aural assertion of individualism. First, otherwise disruptive civilian gunfire became co-opted into the social order by being increasingly militarized. While British troop numbers in Australia gradually declined between the cessation of transportation and the permanent withdrawal of all British forces in 1870, the British tradition of citizen soldiers remained and was, indeed, enhanced. The first part-time volunteer corps since the early colonial years was founded in 1854 and, by the 1860s, most suburbs and towns supported a volunteer unit, chiefly a rifle corps. Apart from drilling, the volunteers mainly practised shooting. Firing a set amount of ammunition at a target across the year served as a rough guide to individual attainment with firearms.[44] Though numbers fluctuated across the next half century, at the height of enlistments, in the 1860s and at the end of the nineteenth century, one in every 50 male workers belonged to a volunteer corps.[45] For these men, who were paid for enlisting, gunfire in this context represented economic remuneration and a certain degree of social standing that came from a recognition of this service to the community. In the larger society, this gunfire constituted affirmative audible evidence of the state's role as protector and of colonial preparedness to assume, when required, the responsibilities of nationhood.

To the extent that gunfire continued as an affirmative language among groups of colonial Australians, it was still heard as a marker of festive celebration but also, and increasingly, as the sound of sport, an association encouraged by the development of breech-loading rifles in the 1860s. But where Sunday shooting legislation was widely heard as targeting the poor, the vitality of the sport of shooting was often linked to its association with privilege. The English sporting magazines that began, mid century, to circulate in Australia reproduced English commentaries that shamelessly spruiked the idea of the gentleman shooter as the personification of 'true sportsman,' a man to whom the 'delights of moderation and prudence' served as 'constant handmaidens and companions.'[46] In this context, gunfire became incorporated into the very model of civility itself.

Along the Cooks River valley, the pastime of shooting became associated with 'gentlemen' as much as tradesmen. Both engaged in well-publicized and well-attended pigeon shooting contests.[47] But, on Sundays, the flouting of the regulations was increasingly associated with better-off citizens who evidently maintained their right to make whatever sounds they chose. In 1896 a member of the Legislative Assembly proposed an amendment to the Sunday Shooting Bill. This prompted a bitter letter to the press in which the writer alleged that not only was Sunday shooting 'very prevalent' but 'especially among persons holding high positions.' Leaving the city on Saturday evenings, such people were able to 'blaze away all day Sunday without fear of interruption from the police, but to the great destruction of game, both in and out of season.' People 'whom one would expect to set a better example to the rest of the community' could simply afford to 'get out of range of the police and baffle detection.'[48] The letter was signed 'A Sportsman.' Eleven years later, in August 1907, in Queensland, the Barcaldine Rifle Club decided, for the convenience of some members, to shoot on Sundays. The captain of the rifle team was Mr. Desgrand, JP, a magistrate. Police arrived to order that they desist, on the Lord's Day, from shooting. When the riflemen sought Desgrand's opinion on whether they should continue, their captain replied, 'Shoot, of course.' Crack went the rifles. That ended the affair. The sergeant acknowledged that the act, 'though spasmodically enforced in the settled districts,' was, like many ancient laws, 'more honoured in the breach than in the observance.'[49] In 1912, another commentator outside of Sydney observed, in relation to Sunday shooting, that in 'these advanced times,' the police turned 'the other ear.'[50] On the acoustic frontier as much as in the city, some sounds were evidently heard with greater clarity.

Conclusion

In contemporary Australia, the sound of gunfire is most immediately associated with war, dispossession, social dysfunction and individual violence. In early colonial Australia, gunfire, as ceremonial sound, inevitably voiced the empire and state as guardian and protector. Within this context, such gunfire also sounded the 'last seedling of . . . Augustan civilization,' a civilization represented by devotion to 'order, regularity and restraint.'[51] But the ritual use of gunfire within the colonies also came to function, on some occasions, as an acoustic safety valve, a release from everyday expectations of order and control. In its carnivalesque uses, gunfire was also the acoustic of those for whom disruption, spontaneous noise at any time, was pleasure. Beyond its celebratory functions, the reports of firearms also sounded a culture fraught with tension: fear of reproducing sectarian division, anxiety over the relationship between the sacred and the everyday and conflict over and within classes over identity, rights and privilege. In the second half of the century were heard, emerging, other means of containing gunfire's potential associations with individualism. There are, of course, many approaches to measuring such movements in history. This study suggests how listening to different strains within one sound might enrich our understanding of the ways in which social relations and common civilities were shaped and renegotiated.

Notes

1 *Sydney Gazette*, 10 January 1818, p. 4.
2 *Historical records of New South Wales*, vol. III, p. 528.
3 Cannon 2004.
4 *Colonial Times* (Hobart), 13 January 1846, p. 3.
5 *Sydney Gazette*, 1 February 1822, p. 3.
6 *Argus*, 14 November 1850, p. 2.
7 *Australasian Chronicle*, 21 January 1840, p. 3.
8 See *Historical records of New South Wales*, vol. VII, pp. 328, 404.

9 *Sydney Gazette and New South Wales Advertiser*, 12 July 1822, pp. 2–3.

10 *Sydney Herald*, 11 December 1834, p. 2.

11 *Perth Gazette and Western Australian Journal*, 8 June 1833, p. 90.

12 *South Australian Gazette and Colonial Register*, 8 December 1838, p. 2; *Hobart Town Gazette*, 25 February 1826, p. 2; *Launceston Advertiser*, 27 July 1829, p. 3.

13 *Australasian Chronicle*, 17 October 1840, p. 2.

14 *Sydney Herald*, 22 October 1840, pp. 2–3.

15 See, for example, Cameron 2006; Frame 2007, p. 63.

16 Gledhill 1931, p. 9.

17 Hassall 1902, p. 67.

18 See, for example, Gledhill 1958, no pagination; *Australian*, 21 November 1839, p. 3.

19 *Sydney Herald*, 22 October 1840, pp. 2–3.

20 *Sydney Gazette and New South Wales Advertiser*, 20 October 1840, pp. 2–3.

21 Shaw 1978, p. 53.

22 *Sydney Monitor and Commercial Advertiser*, 17 October 1840, p. 2.

23 *Australasian Chronicle*, 17 October 1840, p. 2.

24 *Australasian Chronicle*, 10 June 1841, p. 2.

25 New South Wales Legislative Council 1841.

26 *Lectures on the Sabbath* 1841.

27 New South Wales Legislative Council 1841, *Votes and proceedings*, no. 13, p. 10 August.

28 *Geelong Advertiser*, 3 July 1841, p. 2.

29 *Australasian Chronicle*, 28 August 1841, p. 2.

30 *Sydney Gazette*, 24 June 1841, p. 2.

31 Sir George Gipps to Lord John Russell, 16/10/1841, Despatch no. 209, *Historical Records of Australia* 1924, p. 555.

32 'News from an exiled contributor', *Blackwood's Edinburgh Magazine*, vol. LV, no. CCXL, February 1844, p. 195.

33 *Sydney Morning Herald*, 18 November 1845, p. 2.

34 *Geelong Advertiser and Squatters' Advocate*, 14 September 1847, p. 2.

35 *Morning Chronicle*, 12 March 1845, p. 3.

36 Connor 2008, p. 218.

37 Hall 1974, p. 92.

38 Collins 2007.

39 ibid., p. 12.

40 See, for example, W.S. Jevons to Lucy Jevons, 9 April 1859, in Collison Black and Könekamp 1973, p. 370.

41 Clacy 1963, p. 19.

42 See, for example, *Argus*, 20 July 1852, p. 3; 29 September 1852, p. 5.

43 See, for example, Collins 2007, pp. 14–15.

44 Wilcox 1998, p. 19.

45 Ibid., pp. 18, 26.

46 *Bell's Life in Sydney and Sporting Reviewer*, 13 October 1849, p. 2.

47 *Australian Town and Country Journal*, 1 September 1877, p. 3; see also *Bell's Life in Sydney and Sporting Chronicle*, 11 August 1866, p. 2; 6 October 1866, p. 3.

48 *Sydney Morning Herald*, 22 August 1896, p. 10.

49 *Western Champion and General Advertiser for the Central-Western Districts*, 31 August 1907, p. 7.

50 *Singleton Argus*, 10 October 1912, p. 3.

51 Denholm 1979, pp. 30–31.

References

Argus, 1850, 1852.
Australasian Chronicle, 1840, 1841.
Australian, 1839.
Australian Town and Country Journal, 1877.
Bell's Life in Sydney and Sporting Chronicle, 1866.
Bell's Life in Sydney and Sporting Reviewer, 1849.
Blackwood's Edinburgh Magazine, 1844.
Cameron, N. M., 2006. 'Music of the Anglican churches of the Diocese of Sydney, 1836–1868', Master of Music (Musicology) diss., Sydney Conservatorium of Music, University of Sydney.
Cannon, J., 2004. 'George III (1738–1820)', *Oxford dictionary of national biography*, Oxford: Oxford University Press; online edn, May 2013, available from http://www.oxforddnb.com.ezproxy.sl.nsw.gov.au/view/article/10540 [accessed 22 July 2015].
Clacy, Mrs. C., 1963. *A lady's visit to the gold diggings of Australia in 1852–53*, Melbourne: Lansdowne Press.
Collins, D., 2007. 'A "roaring decade": Listening to the Australian gold-fields', in J. Damousi and D. Deacon, eds, *Talking and listening to the age of Modernity: Essays in the history of sound*, Canberra: ANU E Press, pp. 7–18.
Collison Black, R. D. and Konekamp, R., eds, 1973. *Papers and correspondence of William Stanley Jevons*, vol. 2, Correspondence 1850–1862, London: Macmillan.
Colonial Times (Hobart), 1846.
Connor, J., 2008. 'Frontier wars', in P. Dennis, J. Grey, E. Morris, R. Prior and J. Bou, eds, *The Oxford companion to Australian military history*, Melbourne: Oxford University Press, pp. 216–21.
Denholm, D., 1979. *The colonial Australians*, Melbourne: Penguin.
Frame, T., 2007. *Anglicans in Australia*, Sydney: UNSW Press.
Geelong Advertiser and Squatters' Advocate, 1847.
Gledhill, P. W., 1931. 'A venerable church', *Sydney Morning Herald*, 23 May.
Gledhill, P. W., 1958. *A history of St. Peters Church of England Cook's River 1838–1958*, Sydney: Edgar Bragg.
Hall, C., 1974. *Guns in Australia*, Sydney: Paul Hamlyn.
Hassall, J. S., 1902. *In old Australia: Records and reminiscences*, Brisbane: R. S. Hews & Co.
Historical records of Australia, 1924. Series I, Governor's Dispatches to and from England, vol. XXI, October 1840–March 1842, ed. F. Watson, Sydney: Committee of Commonwealth Parliament Library.
Historical records of New South Wales, 1895. vol. III, Hunter, 1796–1799, ed. F. M. Bladen, Sydney: Government Printer.
Historical records of New South Wales, 1901. vol. VII, Bligh and Macquarie, 1809, 1810, 1811, ed. F. M. Bladen, Sydney: Government Printer.
Hobart Town Gazette, 1826.
Launceston Advertiser, 1829.
Lectures on the Sabbath: Occasioned by a recent discussion in the Legislative Council of New South Wales and delivered in the Scots Church, Sydney on Sabbath, July 18th, 1841 and the four following Sabbaths by Ministers of the Presbyterian Church, 1841, Sydney: James Reading.
Morning Chronicle, 1845.
New South Wales Legislative Council, 1841. 'Report from the committee on the shooting on Sunday prevention bill with the minutes of evidence', *Votes and Proceedings*, 27 July.

New South Wales Legislative Council, 1841. *Votes and proceedings,* 10 August.

Perth Gazette and Western Australian Journal, 1833.

Shaw, G. P., 1978. *Patriarch and patriot, William Grant Broughton 1788–1853, Colonial statesman and ecclesiastic,* Melbourne: Melbourne University Press.

Singleton Argus, 1912.

Sydney Gazette, 1818, 1841.

Sydney Gazette and New South Wales Advertiser, 1822, 1840.

Sydney Herald, 1840.

Sydney Monitor and Commercial Advertiser, 1840.

Sydney Morning Herald, 1845, 1896.

Western Champion and General Advertiser for the Central-Western Districts, 1907.

Wilcox, C., 1998. *For hearths and homes: Citizen soldiering in Australia 1854–1945,* Sydney: Allen & Unwin.

8 Sounds and Silence of War

Dresden and Paris during World War II

Joy Damousi

In February 2015, the seventieth anniversary of the bombing of Dresden (13–15 February 1945) was commemorated with great solemnity around the world. Public mourning and official recognition marked the commemoration of the event by German and British governments. In deeply moving and emotional scenes of survivors and their families, as well as sombre speeches by political leaders, there was widespread acknowledgement of the brutality and violence unleashed by this event that has since become one of the iconic symbols of the tragic destruction of World War II.[1]

Military historians have extensively examined the impact of the bombings in many detailed accounts; the event has been identified as one of the key turning points in the course of the war. It has remained controversial for the bombing of a city, which was aimed specifically at women, children and the elderly rather than military targets. Over time, aspects of the narrative about the event have highlighted the German victims of the war.[2] More broadly, as events in military history have increasingly become the focus of examination by cultural and social historians, war has been re-interpreted as a cultural and not exclusively a military phenomenon.[3] In this methodological shift, the experiences of war by individuals, including how they remembered the bombings, have recently come to frame historical accounts of war. It is within this framework that I seek to analyse the impact and centrality of sound in survivor accounts of the bombings.[4] Based on interviews undertaken in Dresden in 2015,[5] this chapter seeks to examine the enduring nature of sound in the memory of the event and the emotional connection to the bombings through sound. The nexus between memory, sound and war is powerfully illustrated in these interviews, and the intersection of these three concepts informs the discussion of the bombings.

In contrast, an examination of Paris under Nazi occupation (May 1940–December 1944) reveals the connection between sound and war through a dramatically different perspective.[6] Sound became a mechanism through which the terror of war was directly experienced on a daily basis for many who remained in Paris after its occupation. Terror was embedded in the sensory and sonic environment in ways that marked a city under totalitarian occupation. One of the striking features of the occupation was that it was not only the threat of military weaponry and overt exertion of violence that controlled the population. It was also the sounds of speech, speaking and listening—in the form of denunciations and eavesdropping—that could have devastating and fatal consequences for civilians.

There are three arguments that underpin this study of these two cities under the conditions of total war. The first is that sound and violence is central to the experience of *terror* in two cities where individuals absorbed its impact in contrasting and complex ways. Second, these accounts further demonstrate that sound is central to how these events are *remembered*. At times specific details are lost, but the description of sound is retold here in ways that are graphic, detailed and often eloquent and powerful. In the disorientation created by war and as physical surroundings collapse and the visual offers no point of reference, it is the fixed nature of sound that often remains in survivor accounts. The clarity of the memory of the war sounds is seemingly never diminished over time. Finally, the richness of these memories provides another layer to further enhance our understanding of experiencing war at an *everyday*, even domestic, level. Sound was embedded in war experience in ways that made it all-pervasive and inescapable.

Dresden, February 1945

The inhabitants of Dresden experienced saturation bombing and with it came thunderous sounds. Dresden was bombed in four attacks between 13 and 15 February 1945 by American and British Allied forces in air raids that decimated the city and left deaths estimated anywhere between 35,000 and 400,000 and with casualties of around 135,000. The heavy bombing of this city illustrates how air warfare had become the supreme form of destruction during World War II. The first attack, by the British Air Force, took place on 13 February 1945, and involved 243 bombers dropping high-explosive and incendiary bombs around a target point in the centre of the city, producing an intense firestorm. The second attack took place several hours later, on 14 February, when a larger group of bombers—529 planes—dropped 1181 metric tonnes of incendiary devices. The third attack took place later on the 14th, when 311 planes dropped 771 tonnes of explosives and firebombs. Finally, on 15 February, 210 American B-17 bombers released 463 tonnes of explosive bombs.[7] A distinctive feature of these attacks was the fires, which erupted from incendiary bombing, and, in particular, the noise and sound of fire, which has dominated many recollections of this incident.

Victor Klemperer, a survivor of the bombing of Dresden, records in his diary how there were 'Big explosions. . . . Bangs as light as day, explosions. . . . There seemed to be no more bombs exploding here, but all around everything was ablaze. I could not make out any details; I saw only flames everywhere, heard the noise of the fire and the storm, felt terribly exhausted inside.'[8] Scholarly attention has been directed towards various aspects of the bombing of Dresden,[9] including the ways in which it has been remembered.[10] This chapter contributes to this discussion by investigating the vital part that sound, generated by bombing, has played in the memory of those who experienced it. In exploring this theme through a range of interviews, the aim is to further contribute to understanding the full auditory experience of this violent event and its enduring legacy through the memories of those who survived it.

Dieter Haufe

Dieter began his interview[11] about the bombings by recollecting an earlier event that had seared into his memory. This involved the burning of a synagogue in 1938. Born in 1933, the inauspicious year in which the Nazis came to power, Dieter was a boy of five years old when he witnessed a moment that would haunt him forever. It would also provide an inexplicable link to the extraordinary events of February 1945. Although the two events—one in 1938, the other in 1945—were seven years apart, they merged in his memory, as they did indeed historically. What drew them together in his collapsed memory? It is not only the continuity of Nazi atrocity, which these dates signify, but also the enduring power of the piercing sounds of both violent events. In 1938, he witnessed the burning of the Dresden synagogue with his father, then a member of the Social Democrat Party, which was eventually banned and members of which were persecuted by the Nazis. Dieter recalled that what he saw, perched on the shoulders of his father, aroused a deep fear in him through the relentless sounds of the hissing flames, which were witnessed by mute, silent bystanders:

> And this [image] comes back to me from time to time: a silent black mass of people. And between them you could see the fire, the fire was hissing, and so [to] say it like a child, [it was] as if a dragon hissed and spewed fire. This is how I kept it in my memory. And when there was a fire, this hissing, like a firestorm, I would find that scary . . . this black mass of people, mute, and the hissing fire. This has been engraved on my mind.

The Dresden synagogue was burned during Kristallnacht on 9 November 1938. By then, the Jewish population of Dresden was in rapid decline, with a drop from 4675 in 1933 to 1265 in 1941. After 1941, only a handful remained. Between 1941 and 1945 they too were sent to the places to which others had been brutally removed—to the camps of Riga, Auschwitz and Theresienstadt. On 16 February 1945, the last of the remaining Jewish population was to be sent to the death camps, but the events of 13 February suddenly halted any further movement to the camps.[12] On that fateful day, the sounds Dieter heard immediately dictated his actions, striking terror:

> Well, we went home, it was lunchtime. On our way from the garden, pre-alarm was sounded. This is a strange sound: *hoo-hoo* [moving up and down], it sounds a little acrimonious. It makes you go faster, automatically. We had to go up to the 4th floor. We had to grab our things and go down to the cellar. We just managed to do that, because when we were up on the 4th floor, full alarm was sounded. Full alarm means a continuous siren tone.

The sirens were a familiar part of everyday life. What made this experience distinctive however was the sound of the planes. Quickly, Dieter, his mother and others in the apartment block took refuge in the cellar. These were rudimentary structures: 'They had put in these wooden pillars and called it air raid shelter.' His mother then decided to go upstairs to retrieve something. Already feeling nervous, his mother's absence struck fear and heightened anxiety into 12-year-old Dieter:

> I wasn't so happy with that, and was a little scared. She went up. And then the sounds came closer. Humming sounds; those were the planes. If I remember correctly, there were 27 planes, so significantly less than in the following attacks. The flak was shooting, and then a new sound was added. They threw detonating bombs. It was like a clicking somehow. But still quite far away because the cellar was quite insulated. My mum was already busy upstairs, when the light started to flicker. And the crashing came closer, and then there was a huge crash. And then there was silence, complete stillness. It was dark; there was no light. We had dust in our mouths. The first thing I did was to take the helmet off and put the gas mask on. They had gas masks for children, they were adapted [from adult ones]. . . . And then I sat there and thought that I would never see my mum again. Something must have happened. We had openings in the walls, I saw that later. After I sat there for I don't know how long, everything vibrated although there was no sound. Everything was quiet.

Dieter was attuned to the sounds in the cellar. The silence too, was menacing and terrifying. Unlike sound, he was uncertain what the silence meant. He immediately felt alone. His mother was nowhere to be seen. He was convinced he would never see her again. Suddenly a neighbour appeared, Frau Muller, a ghastly casualty of the bombing attack, who

> was bleeding all over. She had all these splinters. It turns out that she had gone up like my mum, and we had these huge windows in the corridor. Although it was a little turned away from where the bombs fell,

they . . . fell in a row. And she was right in front of such a window and received all the glass splinters. This woman was later operated on for a long time. She had hundreds of such splinters in her face. But the face, I am unable to forget that: the candle [threw] light on a face which had bloody points everywhere. How can I say it? It was clear to me that I would never see my mum again. This thought did not get out of my mind. I couldn't think of anything else. She then took me by the hand and said that we have to get out. . . . I was hanging close to that woman. I was actually in a whole different world.

But once he went out, in the ensuing chaos, he lost Frau Muller and was immediately disoriented. It was difficult to gauge his bearings, as nothing was left of his physical surroundings. He recalls policemen emerging from the building, but they disappeared 'left and right,' not stopping to help him. He was lost. He tried to recover from the violent soundscape and felt the impact of the pressure of the attack on his ears and therefore his hearing. In the visual chaos of war, with no familiar physical site available to him, it was the sound that became the sensory reference point for him:

Everything was littered with rubble, with stones and debris. If I remember correctly, I sat down on a big stone and didn't know what to do. The noise had faded away, the planes were gone. The cannons did not shoot anymore. And there was pressure on the ears; I already had that in the cellar. It was as if you had cotton wool in your ears. I took off my gas mask. And I couldn't walk properly; it was as if I was walking on cotton wool.

The most reassuring sound for this 12-year-old boy came very soon afterwards, through the emotional relief of another sound:

And then I heard a voice calling my name, in a teary manner. I couldn't quite match that voice. I looked a bit and it was my mother! She also came out of that house down there. And she couldn't find her way either, because it looked so different. We had found each other again. We had no words.

Dieter distinguished in his mind between the sound of the sirens and those of the planes. While his mother 'couldn't bear the plane noises,'

I didn't mind these plane noises as much. Propeller planes have this deep humming sound. Sometimes we have such an old plane here, for sightseeing flights over Dresden. A U52 has also been here. For me this is a more pleasant sound than the damned jets. These have marked me. I don't like them. But the worst sound which haunted me over the following years, even decades (and I was even working in the police force, with the traffic police in Dresden), were the siren tests. This was horrible, the worst. I only wanted them to be over.

Dieter's experience of the terror of the Dresden bombings as a young boy is reflected through his description of sound, of his enduring memory of it and how it became his reference point when his physical surroundings collapsed.

Erika Kügler[13]

As well as the penetrating sounds, it was the silence that is often observed and commented on in the memories of survivors. Erika Kügler (nee Remp) had just turned 16. On the fateful day of 13 February, the sirens alerted citizens to move to the cellars, which Erika did, although the temptation to emerge to listen to the approaching planes was too great. This is how she described the scene:

> On the 13th in the evening, alarm was sounded and we got ready and went down to the cellar. We stayed for a short time in the cellar, and then I went outside the door with someone from the house and we heard how it was humming more and more, so planes were approaching Dresden. But they were still quite far away.

It was a visual spectacle too. Erika recalled the way that the British aircraft would drop flares to light up the area for the bombers to drop the bombs. The Germans called these flares 'Christmas trees.'[14] Erika continued that:

> Christmas trees were set, which lightened up all of Dresden. And the humming got stronger, so we quickly went back to the cellar to seek shelter. We knew that it was Dresden's turn. We came to the cellar and the first bombs were thrown. It was terrible. A terrible bang and a detonating bomb had hit the house.

Once the attack was over, penetrating silence prevailed. 'There was a silence,' Erika noted, 'which you can't describe, a quietness; everyone was occupied with himself. . . . Nobody wanted to say anything.' In the streets, the noise continued, although it was not of bombs, but of the roaring sound of flames. 'Outside it was terrible, you could hear the flames, it was a swooshing noise. And there was a horrible wind, a real storm, a firestorm.' This was a terrifying sound, accompanied by illuminated visual images:

> Yes, it was a horrible swooshing and the wind was whistling, the noises were terrifying. We were lucky that we didn't live in the city centre, but a little outside, so we could access the Elbe meadows and stay there for a while. Of course we could see that everything was on fire and illuminated, this was a terrible picture.

The second attack had all the hallmarks of the first: the sound, the fire and individuals' preoccupation with their own safety:

> Then we heard the humming again. This was the second attack. So we fled to a building on Hindenburgufer, it was a different building, the number 37. We got in, and of course there were many people moving into the cellar. There weren't that many [cellars] available. You had to see that you got in. During the second attack we heard these terrible sounds of impacting bombs. You cannot describe it. When it became quiet we went outside, and stayed at the Elbe meadows until the early morning. Somebody from our house had relatives in Dresden Neustadt, on Albertplatz. And there, at Albertplatz, we were surprised by the next attack the next noon. In the house where we stayed . . . as I already said, you can't actually describe how it was. We sat in the cellar downstairs, and the incendiary bombs were whistling, we realised that something was going on, and a few courageous men went up to the roof and threw down the incendiary bombs. How they did it I don't know. But at least they said: 'There is no danger in the house, it is ok.' This was more or less my experience, and it was devastating, one was occupied with oneself, to get oneself into safety somehow. This terrible noise and sound, and the fire, and the wind, and the only thought was to get out somehow.

There was a 'rustling' sound, a sound of 'rippling/percolating, as if something [was] falling apart.'

Erika's experience in the cellar was one that was distinctively silent as deep reflection became:

> Quiet, very quiet. Everybody was occupied with himself. Everybody made himself as small as possible, as if that way the bomb might miss you. And above all, the noise, these terrible sounds. You realised that the house was caving in but you didn't know how far it would go, if you stayed alive, or what would happen to you.

While experiencing the sound of the bombings was deeply disturbing and powerful, Erika was preoccupied with thoughts of escaping, with getting away and not 'getting in touch with it.' On reflection, too, she believed the experiences of the bombings for a young person might have been different from those of someone who was older. The noise was not traumatizing for her, perhaps because she could focus elsewhere, away from the visual. As a young person, she believed the experience was very different. The urgency to escape preoccupied her and planning the next step was uppermost in her mind, rather than dwelling on the destruction surrounding her.

Ria Wurbach[15]

Silence amidst sound is a similar theme recalled by Ria Wurbach. But the sounds of war did not recede in her memory and, as we shall see, remained with her. The anticipation of the bombs through the approaching sounds is what she recalled. As a child, too, she noted the excitement of 'staying up' later than usual:

> And I still hear it as if it was yesterday: 'Attention, attention, bomber planes approaching Dresden.' Again and again; it was constantly repeated. I suppose my mum had done [everything] necessary and got me dressed. She had a little suitcase where she kept all the papers, and woke my grandma. She lived above us; up there in this room. And then we went to the cellar. So we were under this room here, directly next to the door, and sat down. I can still hear it, this sound has engraved itself in my memory, the approaching bombers. It was a consistent humming, much deeper, it has nothing to do with the modern plane sounds. In one continuous pitch, this consistent humming, which became louder. You still didn't really know what it meant. As a child particularly, I thought it was very interesting that I could stay up. And then [the planes] came closer, and the first bombs came crashing. And the thing flying around was the splinter shield. Down there is the window, and this splinter shield hit us. We sat there next to the door and the splinter shield came flying up. . . .

The splinters from the window came flying in, and the family then made their way to the stairs. It was under the stairs that they crouched, in a very narrow corridor. The sounds penetrated their secluded space. They could hide from the direct line of the bombs, but the sound was all-pervasive:

> The impacts were coming closer and closer. They were so strong. It wasn't only loud—you could hear clearly when the bombs were falling, and you could picture when they exploded, if it was close-by or further away. There was a short whistling sound before. Or sometimes it came very quickly, blow on blow, and the foundation walls were vibrating. The lightning flash, which we could perceive through the cellar, was very frightening somehow. But I remember that I didn't cry. There wasn't any talking in general, none at all. My mother and my grandmother usually chatted a lot, so it was not because they didn't like each other. But then, nothing at all.

Stunned into silence, Ria, her mother and her grandmother feared not only for their lives as the sounds of bombs flew overhead, but there was also a desperate fear of the unknown, of what lay ahead. While she was fearful, as a child, she could not measure the full extent of the catastrophe or the danger involved. Her mother physically panicked in the cellar:

> She didn't scream there . . . she trembled. She right out trembled. And her movements, as far as I can remember, were completely hysterical. While normally she always took care of me, she was completely beside herself then. But they didn't talk much.
> [As a child] you did feel scared, but you could not measure the extent of the danger, and I did not realise it. I just always had the urge to get to the furthest corner, as deep as possible into the earth, not outside but possibly somewhere where there are walls.

The legacy of this experience, now 70 years ago, was that Ria remained sensitive to sound, especially to fireworks. The anxiety of the cellar continued to haunt her:

> The fear, the trepidation, the narrowness in this cellar. These anxieties, which cause nightmares, I can't quite describe it. The sound of sirens, and that I am so bothered by it, it always happens during fireworks. When they detonate something somewhere, certain tone frequencies, I just freeze, I am incapable of doing anything, and it is a strange thing. And I know that there are many people like me. I talked to a carer in a retirement home and she said that, when the end is near, they relive the 13th of February, with all the details. And when I hear about clean wars, that only the building is hit, etc., things that are connected to warfare, I have no enthusiasm at all.

The soundscape and its enduring impact loomed large, and she described turning to stone in response to powerful sounds:

> And although they have become less frequent, these dreams are still haunting me and still today I can't see fireworks. I watch it sometimes when there is no way around it. But these explosions, that people take such pleasure in the loud noise, I prefer to stay away as far as possible. This is really hard to bear, as is the sound of the siren. Their sound is different today, not quite the same tone. But whenever I hear this tone or something similar, I turn to stone. Such explosions, I still freeze, if something of this type happens, I can't function anymore. I remember that later we went on a train, and it made similar sounds. Each time I made a big fuss because I didn't want to go on it, not because I was scared of it, but because of the sound.

But trauma and sound is not restricted to violent sounds and their impact. The oppressive silence also loomed large in her anxiety. When war refugees gathered in the train station, it was the peculiar silence in the station that she recalled:

> What impressed me as a child particularly, was the silence, despite the fact that there were so many people. You would assume that children scream, that people murmur, but this wasn't the case. For me, when particularly depressing images come back from the past, it is this mass of people, and this absolute silence. This was somehow really oppressive.

The combination of sound, silence, noise and anxiety is writ large in this account. These memories are infused with the soundscape, the violence that shook the city and left an indelible mark that continues to haunt those who experienced it. The description of the bombings within family narratives continues to be passed down through the sounds of the event, rather than necessarily the visual impact, which was also chaotic and disorienting.

Heidrun Angermann[16]

Born in 1950, Heidrun recalled how the story of the bombing was commemorated and passed on to her generation. It began with her grandmother's memory and story telling:

> On the 13th of February, grandma always commemorated the bombings, at home, in private, with the family. We were told what had happened then. It usually started like that. It was winter, it was getting dark quickly. We sat together after dinner and were allowed to stay up because of the story. Such things were important for kids. She said that like everyone she also believed that Dresden would be spared, that they wouldn't dare. . . . They heard the first impacts, the howling, and also noticed the firestorm when it started. They then went down to the cellar. . . . Some of the bombs must have howled as they fell down. The impacts were more in the city, they didn't hear that so much, the explosions. At some point it was over. I don't think there was a proper all-clear signal anymore.

Sound was central to the memory and to family history. Heidrun reflected that these stories were largely confined to family history and children rarely spoke about it among themselves. Her grandparents continue to react with sensitivity to noise:

> Yes, it was mostly sirens and propeller planes in the times after the war, but it is still present today. It is relatively handy in case something happens. When the sirens go, everyone pays attention. It is a penetrating sound; so these test runs . . . to make sure that they work, they run siren tests. They always winced, thought about it, then thought: just test

sirens, and calmed down again. It is an affair of half a minute, but there is always this wince, even with us.

There remained a tendency for her grandmother and mother to look up at planes, and the anxiety and tension this created:

Yes, and mum did too; they both looked up. 'What is it? What are they doing? Are they flying peacefully?' If they fly high, they fly far, to Prague. If they fly low, they are coming to us. And with the looking up, again, we only heard the comforting words: 'It is nothing, they only fly with passengers.' And so forth. But it spread to us [the children of her generation]. When a propeller plane came, we looked up, to check if it is not doing anything. . . . And you really have to tell yourself: Yes, it is fine. This is what has been left to us. I can even see that with my two brothers; they are also concerned.

Werner Starke[17]

Werner Starke was born in 1936, so he, like Ria, experienced the war as a boy. He recalled several striking features of the bombings. As a young boy, he remembered, he was expected to be brave, which, it is implied, meant he had to be silent, more emotionally restrained and less vocal than girls and women. His father, a schoolteacher, who became a low-ranking officer, hovered as a towering figure of authority in his story. But even if he could not vocalize the fear, it did not mean he didn't experience it. His fear was compounded by the various sounds he heard:

Of course I was scared when the plaster fell from the ceiling, from the heavy tremors, and I also remember the other children screaming in the cellar. As a boy I had to be brave. My father was an officer; he was also down in the cellar during the night raids, and during the raid next morning he was at work, so he wasn't present. In his presence I was not to make a bad impression as a boy.

In contrast, women and girls screamed, showing considerable weakness, it was implied. Most significantly, however, Werner identified a range of sounds that he recalled—a bursting crash, mines, incendiary bombs, detonating bombs—that all created different sounds. The emotions attached to each sound were clear:

The worst sound I remember, apart from those created by people, by neighbours, were the sirens. The stable tone was the pre-alarm, this was called. And then the mounting and descending tone: *hu-huh*, but high-pitched, that was horrible. It really got on your nerves. And then finally, calming, the repeated warning in a steady tone, that was like a relief.

In Werner's memory, there was an association between classical music and murder, which was enduring and lingered:

> It is wonderful music, which still today makes me wince and makes shivers run down my spine. It is the melody, well-chosen by the propaganda ministry, by Goebbels . . . that played on the radio when there was a special announcement about a German military success. . . . So this is what I connect to this horrible war, because I am telling myself today: each time they played it hundreds of people lost their lives.

For Werner, the sounds of the bombings were revived later in life. 'I re-experienced the bomb attack on Dresden in my sleep,' he said. He recalled these sounds, again, when he experienced an earthquake in the Pamir Mountains and immediately remembered the Dresden bombings.

Paris, 1940–1944

If Dresden provides an example of citizens experiencing the thunderous sounds of war firsthand, the experience of the inhabitants of Paris was in stark contrast. As a city that capitulated to the Nazis, the sound of occupied Paris was not of bombs and air raids. For those in the resistance movement, silence, secrecy and subversion were all part of the way the movement operated in France. Any chance noise could expose political or subversive activity and risk death. Gisele Guillemot recalled how 'A suspicious noise on the staircase and I would think I was about to be arrested, taken to the Gestapo and tortured.'[18]

But overt sounds of the brutal violence of war were never far away. On 16 July 1942, French police arrested 12,884 Jews (including 4501 children and 5802 women) in Paris during what became known as *La Grande Rafle* ('the big round-up'). Most were temporarily interned in a sports stadium, in conditions witnessed by a Paris lawyer, Georges Wellers: 'All those wretched people lived five horrifying days in the enormous interior filled with deafening noise . . . among the screams and cries of people who had gone mad, or the injured who tried to kill themselves.'[19] Within days, detainees were being sent to Germany in cattle-wagons, and some became the first Jews to die in the gas chambers at Auschwitz.[20]

The soundscape of occupied Paris had many layers and levels. Memoirs and contemporary accounts, as well as police records, reflect the importance of silence in order to survive; making *no noise* was imperative. There was eavesdropping and denunciation, which as the police records show, could result in dire consequences. The sound of speech could be a matter of life or death, and Parisians very soon became aware of the danger of 'idle' talk. Radio, as we will see, became central to the lives of Parisians, especially in listening to voices outside of France for political news.

Dominique Jamet recalled childhood illusions of free speech that was in fact anything but free. He summarized the impact of the limits imposed by the occupation on the freedom of speaking and the terror that inflected speech:

> The queues in front of the shop turned into open-air meetings and the virulence of what was being said gave an impression of a complete freedom of speech. This was, of course, an illusion, but how should a non-initiated, a child, guess that what was being said in the shadows of the belltower of Notre-Dame-des-Champs was not necessarily representative of what was thought throughout France, and that the most elementary caution imposed on those who wanted to say something more discordant was to keep their tongues tied. Here, as elsewhere, people knew exactly how much could cost a disobedience to the black-out and the curfew, a banned radio listened to in secret, a rebellious remark, an impertinent comment, a wrong word. An attentive and discerning observer would be surprised that there are so few people who speak, who hold forth, who vituperate, in the midst of these curbed spines, lowered heads. But even silence was often not considered enough of a protection.[21]

The silence indeed could be debilitating. Berthe Auroy described how in 1943 the fear of denunciation was oppressive. 'People now queue in silence,' she noted. 'This general mistrust has a paralysing effect on me.'[22] The Swiss journalist, Edmond Dubois, described in 1946 the way the occupation transformed what and how people spoke in public and the sounds of their voices. People would speak in 'hushed voices,' he observed.[23] Very soon,

> Every Parisian felt the presence of [the] German police around him. Conscious of the danger, he is wary and keeps his mouth shut. The talkative one has stopped offering his opinions in the street, in the metro, in the queues, at the doorsteps of the shops . . . even in private gatherings. He was wary of the telephone.[24]

Even the animals stopped vocalizing. Writing in 1944, the writer Georges Duhamel conveyed how a friend of his who lived outside of Paris observed how a striking feature of the landscape in June–July 1940 was that animals too stopped making their usual sounds. 'Our area is rural,' his friend, Matard said, 'with farms everywhere.' But alarmingly, he observed,

> You could not hear a calf moo, a chicken chitter, a dog bark. In July the animals kept quiet, as if they had understood the terrible misfortune that had befallen France. I can tell you that even the birds in the fields and in the forests didn't sing anymore. And, since the human creatures gave the example, it was a strange concert of silence and sadness.[25]

In the domesticated terror that this soundscape produced, even the most prosaic sounds could arouse deepest anxiety. Hearing the doorbell ring was one. There was fear when the doorbell rang because this might be someone coming to arrest you.[26]

With the exodus of its population, who were fleeing Nazi occupation, Paris was transformed into a silent city. As one report put it:

> The streets are empty at night. Late pedestrians are looking for a house to hide from the first patrols. The city is dark, darker than during the war . . . Paris has gotten younger by 20 years: no motor sounds, and the neighbourhoods further out, such as Muette and Passy, have become the villages again they were 100 years ago. A horse-drawn carriage, some bicycles, from time to time a German military vehicle: this is all in terms of traffic there. The more time passes, the calmer it gets; little by little noisy Paris falls asleep and becomes that place again where the Champs-Elysées were only a little road leading to the countryside.[27]

The silence in the streets was taken up by the sound of boots. Maria Casarès, a French-born Spanish actress, observed in June 1940 how 'the horror of the exodus in France was silent and the streets of the occupation only vibrated from the German boots.'[28] Through a 'a paradox of nature,' observed the lawyer Jacques Isorni, Paris in 1940 'had become more beautiful, a tranquil Paris, without cars, where you could without discovering the slightest sign of the occupation, not hear a single sound of war.'[29] Le Boterf notes how the city became 'mute.'[30]

Denunciation was a new occupation. Marcel Jouhandeau, a collaborationist writer, described his encounter with the Gestapo. In 1943 he was called to the Gestapo headquarters and they asked him about the movements of his best friend:

> His wife has written a denunciation letter!, although he believes that she was not the author. . . . Our [his and his wife's] stormy discussion ended with this conjecture: She must have spoken carelessly about my friends in front of certain untrustworthy people who I detest and who she frequented, because they flattered her vanity, and who had probably used her name without scruple to denounce other people, which they did on a grand scale.[31]

In her published diary of the occupation, Berthe Auroy discussed the new opportunities that arose to make some money through denunciation of fellow citizens. In 1941 she noted:

> There are, it seems, 50 000 (mostly women) traitors to the fatherland who earn 80 francs per day eavesdropping on their compatriots. The Krauts publish enticing advertisements in the newspapers: 'Well paid

post offered, no special skills needed. Contact . . .' Two women who know a friend of Leontine's were looking for a job and went to the address in the ad, but retreated in shock when they realised that there were Germans.

—'Oh sorry, we got the wrong address.'

—'Not at all, my ladies, come in. You are here for the ad. This is what the job is about: We will give you a sector to be monitored in the trains, metros, queues, etc. You will listen attentively to what is being said around you and when you hear disagreeable things against Germany or the Vichy government, you will notify the culprits to the police. All expenses are reimbursed and you earn 80 francs per day.'

Of course, the two ladies went away. What a pretty profession! But on the other hand, one really has to be careful about what one is saying![32]

This misunderstanding of what these advertised jobs entailed seemed to be a regular occurrence. Pauline Corday, a French woman who fled occupied Paris, recorded how (referring to herself): 'A young unemployed girl saw in autumn 1940 an advertisement in the newspaper, offering a job for 4000 francs per month. She went to that address to discover that she was to "listen" in the queues. Horrified, she quickly answered that she had pains in her legs and could not stand for a long time.'[33]

The radio was a popular medium at the time:

I was given the best proof that London radio is universally listened to by an engineer who assured me, without me asking him, that the meter measuring the use of electricity in the whole Parisian area makes a formidable jump at 9.14 pm, and stays at the same level until 9.30 pm. However, radios don't consume much electricity at all, so very many radios must be switched on at the same time to get noticed.[34]

But this was also related to politics. Auroy documented the sounds of English and American voices through the radio, although the Germans attempted to block this. In 1940 she noted:

With regards to Radio Paris, everybody knows that Radio Paris lies. Radio Paris is German. . . . When the broadcasts from London or from Boston start, the radio transforms itself, due to the Germans, into a drum. Sometimes you can hear a formidable waterfall. It is very hard to discern the words under the frequency jamming of the Germans. Sometimes you have to give up. But is such a reassurance to hear the English or American voices.[35]

Conversing while queuing up for food, listening to the radio and hearing the doorbell: these everyday acts became part of the terror embedded within seemingly mundane and routine activities. But the familiar sounds of war were never too far away. Human screams could be all too close. The archivist and intellectual Charles Braibant noted how in June 1944 a friend of his listened to the raids, and 'when they raided the Jews in the Grande-Armée neighbourhood, Marithé could hear the screams of a 75-year-old lady who they took while kicking her with their boots.'[36] The torture and the killings often took place in the city's central heart. Dubois described the sounds in the neighbourhood of rue de Saussaies, which in 1940–42 was the headquarters of Karl Bomelberg, the head of the Gestapo in France who was responsible for Jewish deportations in France:

> But we, the families of the neighbourhood [of the torture centre in the rue de Saussaies] knew better. At any moment we would jump up when we heard the sounds of cars beeping because they would bring in the arrested. On some nights, strange sounds came from the rue de Saussaies: screams from afar, frightened cries 'Mama, mama'. Other times, the echoes from dry clicking sounds, deafened by the walls, clicks which made you think of guns or revolvers. They tortured, killed right in the heart of Paris.[37]

The soundscape of Paris under occupation was varied, complex and diverse. Most strikingly, these examples point to how the intersection of violence and sound permeated domestic space and in doing so was all-pervasive and inescapable. It created a city of terror.

Conclusion

A focus on sound in war provides an opportunity to explore what Mark Smith has described as the total experience of war through the senses.[38] In this chapter, I have explored how the auditory provides a framework for understanding how war was experienced and was embedded in everyday life. Fundamentally, a focus on sound also directs us to a connection to emotion and to memory, given how all are inextricably linked. In the accounts of the Dresden bombings and soundscape of occupied Paris, a narrative of the history of wars and how they have been remembered through *sound* adds a hitherto unexplored dimension to histories of sound and the enduring legacy of sound in traumatic war experiences.

   A further theme is an examination of an overpowering presence of sound in war that transformed the physical environment. In these accounts, too, we can hear the presence of silence, which often heightened anxiety and terror in war experiences in ways that could be as deafening as the shattering sound of warfare. Sounds seem to become lodged in memories of war in ways in which often the mayhem and chaos of the visual does not allow, or the details of which are often forgotten.

Acknowledgements

I am very grateful to all the interviewees who kindly and generously shared their memories and experiences of the Dresden bombings. I would like to thank Dr Jana Verhoeven for her exemplary research assistance in translating the interview material and undertaking the interviews. I am also very grateful to Paula Hamilton for her advice and assistance with this chapter.

Notes

1 See http://www.telegraph.co.uk/news/worldnews/europe/germany/11412224/ Dresden-bombings-Thousands-form-human-chain-for-70th-anniversary.html and http://www.dw.com/en/dresden-commemorates-wwii-bombing/a-18257037. Accessed 12 December 2015.
2 Ten Dyke 2001, p. 82; Taylor 2004.
3 See, for example, Fuchs 2012.
4 For recent studies of war and sound, see Morat 2014. See in this volume: Møller, 'The sounds of World War I: Cheers, songs, and marching sounds: acoustic mobilization and collective affects at the beginning of World War I'; Morat, 'Listening on the home front: Music and the production of social meaning in German concert halls during World War I' and Mansell, 'The sounds of World War II: The silence of Amsterdam before and during World War II: Ecology, semiotics and politics of urban sound'; Rogg 2015, pp. 377–93.
5 Dr Jana Verhoeven undertook 30 interviews of residents in Dresden in 2015. These semi-structured interviews asked participants to recall their memories and experiences of the bombing of Dresden. The interviews were conducted in German and then translated into English. The English transcripts are in possession of the author. This chapter draws from these interviews.
6 For recent works on the history of Paris under occupation, see Mitchell 2008; Spotts 2008; Riding 2010; Rosbottom 2014.
7 Ten Dyke 2001, pp. 81–82.
8 Klemperer 2001, pp. 407–08.
9 Ibid.; Taylor 2004; Addison and Crang 2006.
10 Biddle 2008, pp. 413–49; Fuchs 2012.
11 Interviewed 24 June 2015. All following quotes in this section come from this interview.
12 Crang 2006, pp. 78–95.
13 Interviewed 18 June 2015. All following quotes in this section come from this interview.
14 Taylor 2004, p. 6.
15 Interviewed 22 June 2015. All following quotes in this section come from this interview.
16 Interview 22 June 2015. All following quotes in this section come from this interview.
17 Interviewed 22 June 2015. All following quotes in this section come from this interview.
18 Guillemot 2009.
19 http://www.bbc.co.uk/history/worldwars/genocide/jewish_deportation_01. shtml. Accessed 6 February 2016.
20 http://www.bbc.co.uk/history/worldwars/genocide/jewish_deportation_01. shtml. Accessed 14 November 2015; see Rosbottom 2014.
21 Jamet 2000, p. 178.

22 Auroy 2008, p. 268.
23 Dubois 1946, p. 116.
24 Ibid., p. 121.
25 Duhamel 1944, p. 176.
26 Dubois 1946, p. 122.
27 de Fricambaut and Vacher. Paris, Ville occupée, p. 1.
28 Casarès 1980, p. 112.
29 Isorni 1984, p. 209.
30 Le Boterf 1997, p. 18.
31 Jouhandeau 1980, p. 190.
32 Auroy 2008, p. 174.
33 Corday 1943, p. 70.
34 Ibid., p. 130.
35 Auroy 2008, p. 132.
36 Braibant 1945, p. 500.
37 Dubois 1946, pp. 117–18.
38 Smith 2015, p. 7.

References

Addison, P. and Crang, J. A., eds., 2006. *Firestorm: The bombing of Dresden 1945*, London: Pimlico.
Auroy, B., 2008. *Jours de Guerre: Ma vie sous l'occupation*, Montrouge: Bayard.
Biddle, T., 2008. 'Dresden 1945: Reality, history, memory', *The Journal of Military History*, 72, April, pp. 413–49.
Braibant, C. M., 1945. *La guerre à Paris (8 Nov. 1942–27 Août 1944)*, Paris: Corrêa.
Casarès, M., 1980. *Résidente privilégiée*, Paris: Fayard.
Corday, P., 1943. *J'ai vécu dans Paris occupé*, Montréal: Éditions de l'Arbre.
Crang, J. A., 2006. 'Victor Klemperer's Dresden', in P. Addison and J. Crang, eds., *Firestorm: The bombing of Dresden*, London, Pimlico, pp. 78–95.
de Fricambaut, F. and Vacher, R., 1940–44. *Paris, Ville occupée: [Signé : F. De Fricambault.—J'étais À Paris En Janvier Dernier. Signé: Robert Vacher]*, New Delhi: Bureau d'information de la France combattante.
Deutsche Welle, 2015. 'Dresden commemorates WWII bombing', *Deutsche Welle*, 13 February. Available from: http://www.dw.com/en/dresden-commemorates-wwii-bombing/a-18257037. Accessed 12 December 2015.
Dubois, E., 1946. *Paris sans lumière, 1939–1945; Témoignages*, Lausanne: Payot.
Duhamel, G., 1944. *Chronique des saisons amères*, Paris: Hartmann.
Fuchs, A., 2012. *After the Dresden Bombing: Pathways to memory, 1945 to the present*, London: Palgrave Macmillan.
Guillemot, G., 2009. *From Résistante (Michael Lafon)*, L. Davies, trans., Paris: Broche.
Isorni, J., 1984. *Mémoires, 1, 1*. Paris: R. Laffont.
Jamet, D., 2000. *Un petit Parisien, 1941–1945*, Paris: Flammarion.
Jouhandeau, M., 1980. *Journal sous l'occupation, Suivi de la courbe de nos angoisses*, Paris: Gallimard.
Klemperer, V., 2001. *I will bear witness: A diary of the Nazi Years 1942–1945*, New York: The Modern Library.
Krol, C. and APTN, 2015. 'Dresden bombings: Thousands form human chain for 70th anniversary', *The Telegraph*, 13 February. Available from: http://www.telegraph.co.uk/news/worldnews/europe/germany/11412224/Dresden-bombings-Thousands-form-human-chain-for-70th-anniversary.html. Accessed 12 December 2015.
Le Boterf, H., 1997. *La vie Parisienne sous l'occupation*, Paris: Éditions France-Empire.

Mansell, J., 2014. 'The sounds of World War II: The silence of Amsterdam before and during World War II: Ecology, semiotics and politics of urban sound', in D. Morat, ed., *Sounds of modern history: Auditory cultures in 19th and 20th century Europe*, New York: Berghahn Books, pp. 305–24.

Mitchell, A., 2008. *Nazi Paris: The history of an occupation, 1940–1944*, New York: Berghahn Books.

Møller, S. O., 2014. 'The sounds of World War I: Cheers, songs, and marching sounds: Acoustic mobilization and collective affects at the beginning of World War I', in D. Morat, ed., *Sounds of modern history: Auditory cultures in 19th and 20th century Europe*, New York: Berghahn Books, pp. 177–200.

Morat, D., 2014. 'Listening on the home front: Music and the production of social meaning in German concert halls during World War I', in D. Morat, ed., *Sounds of modern history: Auditory cultures in 19th and 20th century Europe*, New York: Berghahn Books, pp. 201–24.

Morat, D., ed., 2014. *Sounds of modern history: Auditory cultures in 19th and 20th century Europe*, New York: Berghahn Books.

Riding, A., 2010. *And the show went on: Cultural life in Nazi-occupied Paris*, New York: Knopf.

Rogg, M., 2015. 'Lauter Krieg—Annäherung an eine Militärgeschichte als Klanggeschichte', in M. Jonas, U. Lappenküper and O. von Wrochem, eds., *Dynamiken der Gewalt-Krieg im Spannungsfeld von Politik, Ideologie und Gesellschaft*, Paderborn: Ferdinand Schoeningh, pp. 377–93.

Rosbottom, R. C., 2014. *When Paris went dark: The city of light under German occupation, 1940–1944*, New York: Little, Brown and Company.

Smith, M. M., 2015. *The smell of battle, the taste of siege: A sensory history of the Civil War*, Oxford: Oxford University Press.

Spotts, F., 2008. *The shameful peace: How French artists and intellectuals survived the Nazi occupation*, New Haven, NY: Yale University Press.

Taylor, F., 2004. *Dresden: Tuesday February 14th 1945*, London: Bloomsbury.

Ten Dyke, E. A., 2001. *Dresden: Paradoxes of memory in history*, London: Routledge.

Webster, P., 2011. 'The Vichy policy on Jewish deportation', *BBC History*, 17 February. Available from: http://www.bbc.co.uk/history/worldwars/genocide/jewish_deportation_01.shtml. Accessed 14 November 2015.

9 Hearing the 1965–66 Indonesian Anti-Communist Repression

Sensory History and Its Possibilities

Vannessa Hearman

Rrrrem, rreeem, reeeem. The sound of the trucks was like the sound of death to the prisoners. Anyone loaded on to the trucks and taken away would never return. The prisoners heard stories that those taken away were murdered. They were taken blindfolded to Babau (sic), the district centre, then again by truck to the place of slaughter which was already prepared.[1]

Sound forms the backbone of farm labourer Niko's memories of the 1965–66 anti-communist repression in Indonesia, as he related above in his interview for an oral history collection, edited by fellow former political prisoner, Putu Oka Sukanta. From Southeast Sulawesi, Niko was among those persecuted following a coup attempt in Jakarta that was blamed on the Indonesian Communist Party (Partai Komunis Indonesia, PKI). Half a million people were killed and hundreds of thousands imprisoned in the anti-communist purges. This chapter is a preliminary exploration of the soundscape of this period of violence. It was first motivated by my experiences of interviewing survivors of the purges, in which they related how their memories of certain sounds from their period of imprisonment were particularly potent. Episodes of dramatic social change such as revolution are unlikely to be quiet affairs. As David Hendy writes, 'For those caught in the thick of the upheaval and violence, the experience might even be defined by noise more than anything else; what they remember, even years afterwards, is the visceral shock or disorientation of overwhelming and unceasing din, the terror or the foreboding at the danger it foretells, *the sheer auditory theatre* of it all' [emphasis added].[2] In Indonesia, the purges and the regime change they ushered in were marked with a soundscape that was frightening and long lasting, particularly for the victims.

In constructing a sensory history of the repression, I aim to examine the social and cultural construction of hearing and 'its role in texturing the past.'[3] How might the repression have sounded? What evidence do we have? Drawing on interviews I conducted in Indonesia and a range of oral history collections, I reconstruct the sounds of repression of the period, as well as posit some arguments on why the sharpest memories about the time of violence are those that relate to sound. Historians Alain Corbin and Mark Smith point to the restriction of available sources for sensory historians, including the 'transience of evidence,' which makes the study of habitus necessary in order to approximate reconstruction of the past.[4] Smith questions the extent that past sensations are reproducible for present day audiences.[5] As well as restrictions on available sources, human perceptions about their surroundings have changed over time, and the established hierarchy of the senses is time-bound. For example, traffic noise in the city is no longer remarked upon or recorded, in contrast to previously, hence the need to be sensitive to context and habitus.[6] Furthermore, print continues to be an inescapable medium for historians in terms of available sources and how we in turn make available the histories that we produce.[7] Despite these perceived limitations, sensory history 'encourages the adoption of a comprehensive viewpoint.'[8] A sensory approach has not been attempted before in the context of Indonesia, although studies that draw on sound and the other senses can be found in disciplines such as art history, film and performance studies.[9]

While oral histories and memoirs about the Indonesian violence are replete with sensory recollections,[10] there has been no scholarly analysis using the senses as a way to analyse the experience of violence. The violence was felt acutely, bodily and sensorially, and a careful sifting through testimonies and oral history collections attests to this fact. The sensory landscape of the violence is reflected in some literature through the descriptions of the treatment of political prisoners and the use of allegories related to sound.[11] The purges lend themselves to a sensory approach, being the visceral transformation that they represented to Indonesia that was experienced fully through the senses. It is also useful for understanding a further layer of the daily routine of violence and the centrality of sound to the everyday experience of violence.

Indonesia, the Cold War and the slaughter of the left

In 1965, Indonesia had the world's third largest communist party after the Soviet Union and China. Enjoying a close relationship with the communist world at the height of the Cold War, Indonesia under President Sukarno was a cause for concern to the United States and other Western countries. They feared that the strategic nation would fall to communism. Following independence from the Dutch in 1949, Indonesians were involved in various forms of social and political activism, in professional and sectoral organizations and political parties. Under Sukarno, Indonesia was highly critical of the Cold War division of the world into two opposing camps and sympathetic to the plight of newly independent nations caught up in this conflict. However, not all in Indonesia shared Sukarno's political agenda of implementing an Indonesian style of socialism with the support of the

PKI, and opposition increased after he introduced Guided Democracy from 1959 onwards. Under this system, multi-party parliamentary democracy was suspended in preference for a body of representative groups and increasing the power vested in the president. Those who criticized Sukarno were imprisoned or experienced other forms of sanctions. Elections were postponed indefinitely and the media was subjected to close scrutiny. Under such conditions, those opposed to Sukarno and the PKI, such as religious organizations and parties and sections of the military, formed their own alliances. A great deal of tension resulted between the different political blocs, despite Sukarno imposing an ideological orthodoxy under Guided Democracy.[12]

On 30 September 1965, a group of soldiers and officers calling themselves the Thirtieth September Movement, led by Lieutenant Colonel Untung, kidnapped and killed seven army officers at Lubang Buaya on the outskirts of Jakarta. Those killed included the highest level of the Army leadership. Untung and his men argued their operation was a pre-emptive move to prevent a right wing Army-led coup against President Sukarno. They broadcast on state radio on 1 October that they had formed a Revolutionary Council to replace Sukarno's cabinet. The Movement was unsuccessful in winning broad support inside the military, and in a matter of days they were defeated. The Army under then Major General Suharto used the 'coup attempt' as a pretext to suppress the Indonesian Left as a whole and for the replacement of President Sukarno by the Army-led New Order regime (1966–98).[13]

As the bodies of the Army officers were exhumed on 4 October and newspapers were briefly suspended from publishing, Army newspapers falsely reported that leftist women from the organization Gerwani (Gerakan Wanita Indonesia, Indonesian Women's Movement) had sexually tortured and mutilated them. Furthermore, they reported that communists were setting out to slaughter their opponents throughout Indonesia. A coalition of political parties and civil society organizations, the KAP-Gestapu (Action Committee against the Thirtieth September Movement) arose to mount an effective opposition to the PKI. The KAP-Gestapu held demonstrations in Jakarta criticizing Sukarno and demanding the banning of the PKI in the aftermath of the Thirtieth September Movement.[14]

In early October, anti-PKI killings began in Aceh, in the western part of Sumatra, and were in full swing by December 1965 in Java and Bali.[15] Soldiers, police and civilian militias raided leftist homes and communities. The repression exploited local opposition to the PKI, such as religious organizations or rival political parties that in turn provided active support for the violence. While the repression had subsided by March 1966 in Java, Bali and Sumatra, the 'clean ups' continued on a smaller scale until the late 1960s. On the island of Flores the repression only began in early 1966 and was directed by the military as part of a law and order operation called KOMOP.[16]

In mobilizing the forces to carry out the suppression campaign, instructions were given in a mixture of formats: in letters and telegrams, over the telephone, on the radio and in face-to-face meetings. In the massacres in East Java, civilians, drawn mainly from the Islamic organization Nahdlatul Ulama (NU), worked in tandem with the security forces. Interviewees from religious organizations recalled that they were invited to meetings at local military headquarters, where they were warned of impending communist violence and urged to act decisively, and if necessary, violently, against communists.[17] Muhammad, a local youth leader in the small town of Bangil, spoke of how he heard a public address at the town's cinema by NU leader Subchan ZE, who had come from Jakarta, in which he called on the Sukarno government to disband the PKI.[18] Muhammad's group then heard from a local religious leader that it was acceptable to kill communists in the area as communists had tried to overthrow the government.[19] Utterances from local authorities therefore played an important role in galvanizing young men from religious groups into action in East Java.

Marcus (pseudonym) was a philosophy student who was training to be a Catholic priest in the small town of Kediri. He attended a meeting organized by the local military to convince young religious men to participate in the killings.[20] He had been told of the strong likelihood of communist attacks against the Catholic seminary or the church. The violence in Kediri began on 11 October, when the PKI headquarters came under attack. A curfew was in place until November 1965 in the area, thus limiting the ability of its inhabitants to see the massacres. People remained inside. Sounds in the night were some of the only ways that people recall experiencing the repression. As the violence intensified, the sound of metal on metal, of knives being sharpened outside communist homes or in communist areas, was Marcus' most vivid memory. Marcus was sad that his community had suffered so much in 1965, and the sound of knives being sharpened in an attempt to intimidate communists was his clearest memory of the time.[21]

The method of killing differed according to each locality and to the perpetrators, whether they were civilians or members of the security forces. In Joshua Oppenheimer's first film *The Act of Killing*, perpetrator Anwar Congo re-enacted how he strangled those detained by his North Sumatra gang with a piece of wire.[22] Although perpetrator testimonies are relatively rare, killers have discussed in their testimonies how they dispatched their victims, which included shooting, throat-slitting, beheading and pushing into steep vertical caves.[23] In East Java, NU youth gangs decapitated victims with swords and other sharp weapons. Soldiers also shot victims in Jember, East Java and in Central Java. Victims were then buried in mass graves or disposed of in waterways.[24]

Interviewees, such as Mbah Wiryo, who lived 100 metres from the Bacem Bridge in Surakarta, Central Java, remember the sound of gunshots and objects falling into the water during the killings.[25] People locked their doors at the time of executions, which according to her, occurred every two or three days over a period of six months. Residents dared to leave their houses only once dawn approached. 'If we went outside, we might get accused of being communist,' she said.[26] Detainees were shot on the Bacem Bridge in such a way that their lifeless bodies fell onto the rocks and into the river below. If necessary, their bodies were collected or helped into the river's flow in the morning by specially designated 'volunteers' or local residents. One of these, a man named Bibit, also lived near the bridge and listened to the shots and sound of falling objects; as he recollected:

> The sound of pistol or carbine shots could be heard clearly. Almost every night for two years on and on they went, particularly from 1966 to 1967. We just had to count how many shots were heard each night. Five shots meant five dead. Dozens of shots meant dozens dead.[27]

In West Timor, the Army was responsible for suppressing the communists. Almost every night shots were heard in the town of Soe, and Army cars drove around the town patrolling and picking up detainees, according to a policeman named Beny.[28] Beny was haunted by his involvement in the killings.

While there is little solid evidence that the killings and repression were coordinated across Indonesia, there were many similarities between localities in the way the repression was carried out. In Java, the early stages of the killings involved greater visuality compared to the more routine slaughter that took place in the later stages. Bodies and body parts were displayed in the streets and were floated along the waterways, such as the Brantas River in East Java. As observer Pipit Rochijat recalled, severed penises were displayed outside a prostitution complex in Kediri, East Java, during the killings, as a moral warning to those frequenting the complex and as a warning to leftists.[29] Constance Classen has discussed how visualism plays a dominant role in the modern West. Visualism grew in importance as literacy increased from the eighteenth century onwards.[30] As the printed word grew more accessible, less attention was paid to the other senses as sources of human experience. It is difficult, however, to argue that the focus on sight was confined to the modern West. For example, in the first stage of the killings in Java, the traces of the killings were observable in public places. These displays were designed to shock the population and to normalize violence against the left.

Despite the high degree of visuality in the killings, the tools of the repression and the way it was carried out *limited* the visibility of the repression. Information blackouts in the form of the suspension of newspapers and the circulation of black propaganda regarding the Gerwani women sowed confusion and a growing reliance on rumour. Curfews and roadblocks were in place in many parts of Java and raids against known leftists took place after dark. These measures limited the visibility of the killings, and people sought information by relying on their other senses such as hearing. Unlike vision, the path of sound was difficult to control. Bystanders could still refuse to view remnants of the killings in public places by turning away or avoiding such places until late in the day, by which time corpses were likely to have been cleared by members of the community or the authorities. In contrast, hearing sounds is involuntary, unless one acts to block all noise from coming in. Sound in turn requires the brain's engagement in *thinking* about what one happens to hear. As Hendy has argued, hearing adds a whole layer of immersive experience.[31]

The sound of trucks driving by and pulling up was associated with impending violence. After dark, listening out for the sound of the trucks was one way of determining possible danger. For example, Ling (pseudonym) was a young woman whose father had been involved in the leftist organization Baperki (Badan Permusyawaratan Kewarganegaraan Indonesia, Consultative Body for Indonesian Citizenship) just prior to the repression. Ling was concerned that she would be arrested, as she had performed at Baperki's gatherings the song 'Genjer-genjer.' It was a song that told of the desperation of peasants in Banyuwangi, in the eastern part of the province during the Japanese occupation (1942–45), who resorted to eating water hyacinths. Although it was a nationalist song critical of the Japanese, it became associated with the left and with Gerwani in particular. According to Army propaganda about the Thirtieth September Movement, Gerwani women were supposed to have sung the song while sexually mutilating the Army officers at Lubang Buaya. Feeling fearful, Ling sat with her ears pricked up late into the night in her home village of Batu. She recalled the sound of trucks driving up her street and the sound of boots or shoes hitting the ground. She said, 'My knees shook together till they rattled. I was so scared, always wondering if the trucks would stop outside my door.'[32] Ling listened to the sound of the trucks night after night for several weeks, but the trucks never stopped outside her house. The sounds, however, continue to haunt her.

Prison Soundscape

Ling's fears of being detained were justified, as hundreds of thousands were arrested and held in connection with being involved in the Thirtieth September Movement. Former political prisoner and journalist Oei Hiem Hwie was one of those detained. In an interview in Surabaya, Oei told me that he was seeking a writer who could write his life story. Rather apologetically, he told me that despite being a writer, he was unable to write his own life story.[33] Writing one's life story was at the time of the interview a popular pastime for those who had been imprisoned in connection with the 1965–66 purges.[34] Oei explained that it was not for want of trying. Rather it was because he felt traumatized. Each time he set up his typewriter to write about his prison experiences, he began hearing the sound of keys in the hands of his prison guards. Oei was arrested in Malang, East Java, in 1965 and then imprisoned in several prisons in Surabaya before being transported to the island prison of Buru in the Maluku islands, where 12,000 Javanese men were detained. The sound of keys in his memory of the time inhibited his ability to write. That such a recollection was so difficult to manage for Oei led me to exploring the power of sound and hearing in prisoners' memories.

Prisons, halls, hospitals, factories, schools, hotels and government buildings were all pressed into service to house the detainees. Djoko Sri Moeljono was a metallurgist at the Cilegon heavy industrial plant in West Java. On the day he was arrested, he recalled, 'From a distance, a group of demonstrators could be seen carrying pieces of wood, bamboo and other things, yelling loudly, "finish the PKI, hang [PKI chairman] Aidit." '[35] The sound of crowds therefore heralded the advent of violence. Moeljono was detained for 14 years, including on Buru Island. Like hundreds of thousands of other Indonesians, he was never tried.

The daily routine of detainees was marked with a rich soundscape of army trucks, the prison bell, jingling keys and footsteps of prison guards. Prisoners complained that they found the nights most menacing and frightening in prison in their cells. Moeljono recalled, 'Each night after 6 pm, every hour we heard the prison bell marking time coming from the watchtower in each corner of the prison.'[36] The prison bell acted as the collective organizer in the prison, indicating the passing hours. The French village bells that Alain Corbin made the focus of his study were a feature in the lost world of nineteenth century France, a rural world marked by the peals of the bell 'which were listened to and evaluated according to a system of affects that is now lost to us.'[37] Not unlike the village bells, the prison bell was read alongside other sounds, such as the jingling of keys in guards' hands, to create a certain auditory landscape that was by and large controlled by those in charge of detention.

The disappearance of prisoners from detention centres, usually under cover of darkness, consumed many lives. In a practice called *dibon* or 'being borrowed,' guards took detainees out of their cells in order for them to be interrogated, but they did not always return.[38] Prisoners associated the sounds of jingling keys, footsteps and the opening and closing of doors with interrogations and disappearances of their fellow inmates. They also feared that they would be targeted next. A lecturer in history education, Harsutejo, was imprisoned in Lowokwaru, near Malang in East Java, from the fourth week of October 1965. Prisoners, including prominent local leaders, began to disappear from that prison each night.[39]

> At 2 or 3am people were woken up. Especially on those rainy, December nights. People were told to get onto trucks. They were tied up, then they immediately disappeared. Thousands of people from Lowokwaru. Each day [people were taken away]. Including the owner of newspaper

Trompet Masjarakat [*The People's Bugle*], Goei Po An. His cell was opposite mine. We had the chance to chat a few times. One night he got taken. I don't know. Gone forever.[40]

At the time Harsutejo was not particularly perturbed by the disappearances. But memories of the disappearances made him feel extremely sad and uncomfortable years later. The sound of rain, followed by the jingling of keys and the arrival of prison guards, heralded in his memory the imminent disappearance of his comrades. Similarly, although he was held in a different prison than Harsutejo, far away in West Java, Moeljono (who is also Harsutejo's brother-in-law) outlined how certain sounds signalled imminent torture, interrogation and political disappearance:

> Everybody's hearts would beat faster when at night the guard opened the door connecting the office with the cells. Each person pricked up his ears. When the guard left, leaving in his wake the sound of his bundle of keys, we all sat in silence waiting for the prison bell to sound. Every hour from the guard tower at the corner of the prison, a bell sounded. Sometimes the situation was so tense and frightening, it prevented [the inmates] from sleeping, even until morning when our friends [other inmates] who were 'borrowed' were returned to their rooms.[41]

Held at the Sasono Mulyo Royal Hall in Surakarta, Central Java, another former detainee, Mulyadi, related how disappearances began with the creak of the gate as it opened to allow a guard to enter.[42] Detainees who were relaxed and chatting among themselves suddenly stopped speaking when they heard the guard blowing on the microphone, because soon he would begin to call out the names of those prisoners who were listed for interrogation that night. Prisoners were rendered speechless by the creak of the gate and the blowing on the microphone. In that way, control of the auditory space passed from the men to their jailers.

During interrogation, many, but not all, prisoners were tortured. In his article on torture and the creation of untruths by the New Order regime, John Roosa writes that the interrogators 'targeted the body of the prisoner, employing ancient simple methods' with the 'innocuous, every day objects of the office' becoming instruments of pain.[43] Prisoners were tortured using lit cigarettes burnt onto bare skin. The leg of a chair or table could be placed on detainees' toes or wielded like a club. Whips and stingray tails cut through their flesh. Electrocution was done using a primitive device cranked by hand. Tan Swie Ling was a young ethnic Chinese man who was imprisoned in Jakarta for harbouring the communist leader Sudisman. Detainees occupied every available space at the Gunung Sahari headquarters of the dreaded Operation Bat, a detention centre run by Army intelligence, so Tan had to sleep on the verandah outside the torture room. He deduced how the torture was carried out from the sounds he heard. He recounted:

> I could hear and understand what was going on in the interrogation/ torture room and knew well the cry of pain. I knew how the person inside was being tortured. Slow cries of pain meant he was being whipped with the stingray tail and vibrating hysterical screams were caused by electrocution of the genitalia. Because I became accustomed to hearing people being whipped with stingray tail, I then knew how many lashes of the whip A or B received. In that way, I guessed the weight of the accusations as well as the attitude of the detainee in facing interrogation.[44]

Sound therefore provided him with (unreliable) clues as to a detainee's culpability. The use of torture did not produce truth, but reinforced the regime's version of events and the guilt of communists and the need for them to be punished.[45] In line with Roosa, I argue that the regime's use of torture, in a situation of confusion and little information, did help shape the detainees' perceptions about their fellow detainees, however erroneous.

Tan further highlights the immense power of the senses and the gravity of torture, as he cautioned other detainees against peeping or eavesdropping on torture. He wrote, 'Peeping in or eavesdropping on detainees who were being tortured was no small matter. It was only safe when done by those who had been tortured before, but very dangerous for newcomers who had not been interrogated/tortured.'[46] A new detainee, a young man who was a member of the Communist Youth, arrived in Tan's detention centre from a town in East Java. Burnt by curiosity, the newcomer watched and listened in to interrogations and torture day after day. The terror broke him mentally, as he became fearful of how he himself would survive similar treatment. Images and sounds he received proved too powerful for the young man, according to Tan.

The sound of the army trucks that transported detainees to and from places of detention was also a key part of the auditory landscape, as Niko's opening testimony illustrates. Niko recalled, 'When we heard that sound, our hair would stand on end and we would experience the most terrible fear. Some of the prisoners would be loaded on to the trucks. We would wait for them to come back, but none of them ever did, to this very day.'[47] For Niko and his fellow inmates, the rumble of the truck was 'the rumble of approaching death.' Their fear precluded sleep and they spent almost every night through the month of December with their 'ears listening out for the rumble of the truck.'[48] The auditory landscape of Indonesian prisons and detention centres in the mid-1960s perhaps did not differ markedly from similar instances of political repression elsewhere. The sound of imprisonment had a markedly lasting effect on those being detained because the immersive quality that hearing added to the human experience underscored the power imbalance in which survivors found themselves. As Hendy elucidates further, 'Intimately related to power is the issue of control. By this I mean that degree of control any group of people have had over the soundscape in which they found themselves living at any given time.'[49] Sound reinforced powerlessness, as those who controlled the environment controlled what the detainees heard.

However, those who ran the detention centres did not possess absolute control of the auditory landscape. As Evan Kutzler argues, 'Aural landscapes were sites of interpretive struggles between prisoners and guards.'[50] Listening was a highly subjective exercise, which resulted in 'clashing interpretations of the same resonances.'[51] In detention, people sought out news about the outside world from new arrivals, prison guards and visitors, and by secretly listening to the radio or picking up scraps of newspaper.[52] Detainees exchanged stories.[53] They wrote poetry, music and songs to drive away loneliness and to keep alive a sense of hope. Because of the absence of writing paper and implements, writer Pramoedya Ananta Toer recited to his fellow exiles on Buru Island the historical fiction he was composing, which later became the Buru Quartet.[54] Later, the Command Headquarters Band was formed, initiated by prison authorities and which drew disagreement from some of the detainees, because it was perceived to be a creation of their jailers.[55] Musicians from various prison units on the island were relieved of hard labour and drawn to headquarters. They then toured the island, performing from barrack to barrack. For the detainees, the performances provided relief from the daily grind. Women detainees in Plantungan women's prison in Central Java sang to mark days such as Independence Day on 17 August and formed traditional Javanese singing groups.[56] Detainees also participated in religious choirs that performed at mass in prisons every Sunday.[57] In these small ways, they challenged the silencing imposed upon them and overcame some of the harshness and powerlessness of their situation, in which their jailers controlled the auditory environment and prohibited their rights to information and entertainment.

The worst of the killings had stopped by March 1966, but their impact continues to be felt in the physical environment in which they occurred. As Adrian Vickers writes, 'The physical landscape of Java and Bali are contested, because it is a landscape polluted on many levels.'[58] By this, Vickers means that post-New Order Indonesia is unable to put the dead to rest properly, despite Indonesia now being more able to come to terms with the past after President Suharto resigned in 1998. While their bones are accidentally unearthed from time to time, in the absence of an alternative narrative able to counter the strong New Order narrative in which these dead are ungrievable, Indonesians are in turn unable to put these dead to rest. Attempts to exhume mass graves and rebury the remains found within them have been met with violence.[59] The silencing of victims through killings and imprisonment competed with the sounds that would not be silenced, the sound of hauntings. From Sumatra to Eastern Indonesia, stories of hauntings circulate. They constitute one way of expressing grief and regret about past violence, but are also tied to many traditional beliefs throughout the archipelago, such as the belief that those who lose their lives violently are unable to rest until certain rituals have been performed to placate them or events signifying closure have taken place. Former policeman Beny from Soe, West Timor, believed that killing sites were sacred (*keramat*). According to him, people in the area still heard screams in the night at places of execution near Soe.[60] In Sikka, Flores, at the site of a suspected mass grave, according to another witness, 'Local residents hear strange sounds, like people singing.'[61] Sound-based haunting events are related to the killings being experienced only as distant sounds at the time. Therefore it is not surprising that the hauntings are also based on sound.

The soundscape of the 1965–66 repression in Indonesia is still to be explored, but there are signs that the music of the 1960s, as a marker of history and as a way to convey memories of suffering, is awakening interest among the younger generation. As Steven Farram has shown, Indonesia in the 1960s was a place of heightened nationalist sentiment, and some of this was expressed through music and songs, including in Sukarno's opposition to Western and rock music.[62] In 2014, activists based at the 1965 Park in Bali began to compile onto a compact disc some of the songs from the period, as well as those that detainees wrote and sang. Started by a group of Denpasar youth, the 1965 Park functions simultaneously as a meeting space, reading room and place of reflection and is located in the compound of a family who lost one of its members in the violence.[63] The PKI and Marxism-Leninism are still banned through a 1966 parliamentary decree, so the activists were careful not to choose songs that the government or the military and civilian groups involved in the violence could use to accuse them of trying to revive the PKI or its ideology. The songs chosen included for example 'Si Buyung' ('The Boy'), which is a lament from the composer for the baby son he had yet to meet, as the son was born after the composer went into prison.[64] Former political prisoners supplied the music and lyrics to young activists and musicians who recorded the songs in the studio.

Conclusion

The seizure of power by the Army and the destruction of the Indonesian Communist Party were significant events in Indonesia's history. Hendy's reminder of the gravity of the transfer of power as an auditory event, albeit originally provided in a different context, serves us very well: 'Powerlessness was not some abstract condition but something experienced all the time by people through their own senses.'[65] Sound was central to the terror experienced by the population during the anti-communist violence. It was all-pervasive and ubiquitous and yet its importance has not been fully realized and analysed in studies of the experience of this brutal chapter in Indonesian history.

Towards the end of the interview with him, the interviewer asked Niko, the farm labourer who was haunted by the sound of trucks, 'Are those old trucks still around? And how do you feel when you see them now?' Niko answered, 'Yes, the trucks are still around. When I am at the market and they go past, it reminds me of 1965. But now God helps me, and I am safe.'[66] At long last, Niko has overcome his fear of the trucks. But many others have not been as lucky. The sensory experience of these dramatic events has resulted in decades-long trauma that has rendered them powerless in certain moments, such as in the act of writing one's life story. For historians, a sensory approach yields insights into the relationship between sound and trauma and helps us understand the long-lasting effects of terror from violence that is channelled and experienced through the senses.

Notes

1 Sukanta 2014, p. 323.
2 Hendy 2013, p. 247.
3 Smith 2007, p. 4.
4 Ibid., p. 119 and Corbin 1995, p. 184.
5 Smith 2007, p. 121.
6 Corbin 1995, p. 190.
7 Smith 2007, p. 125.
8 Corbin 1995, p. 183.
9 These include Hatley 2008; Weintraub 2010; Farram 2014, pp. 1–24; Heryanto 2014.
10 See for example Setiawan 2006; Tan 2010, pp. 22–23; Harsutejo 2013; Moeljono 2013; Sukanta 2014.
11 Amnesty 1977, pp. 71–89; Roosa 2008; John Prior 2011.
12 Cribb 2001, p. 231.
13 On the events on 30 September and thereafter, see Roosa 2006, pp. 3–4.
14 Crouch 1988, p. 141.
15 Cribb 2001, p. 233.
16 Prior 2011.
17 Hearman 2012, p. 114.
18 Muhammad (pseudonym), interview, Bangil, 12 February 2009.
19 Ibid.
20 Marcus (pseudonym), interview, Jakarta, 3 February 2009.
21 Ibid.
22 *The act of killing* 2013.
23 Tempo 2013.
24 Cribb 2001, p. 233.
25 Tempo 2013, p. 50.
26 Ibid.
27 Setiawan 2006, p. 232.
28 Sukanta 2014, p. 32.
29 Rochijat 1985.
30 Classen 1993, p. 27.
31 Hendy 2013, p. 382.
32 Ling (pseudonym), pers. comm. 16 February 2009.

33 Oei Hiem Hwie, interview, Surabaya, 2 August 2007.
34 Hearman 2009.
35 Moeljono 2013, p. 28.
36 Ibid., p. 31.
37 Corbin 1998, p. xix.
38 Farid 2005, p. 8.
39 Harsutejo, interview, Jakarta, 30 April 2007.
40 Ibid.
41 Moeljono 2013, pp. 33–34.
42 Mulyadi, interviewed 25 September 2012, in *Jembatan Bacem: Film dokumenter tentang Peristiwa 1965*, 2013.
43 Roosa 2008, p. 40.
44 Tan 2010, pp. 22–23.
45 Roosa 2008, p. 49.
46 Tan 2010, p. 22.
47 Sukanta 2014, p. 323.
48 Ibid.
49 Hendy 2013, p. 384.
50 Kutzler 2014, p. 248.
51 Ibid.
52 Harsutejo 2013, p. 131; Moeljono 2013, pp. 39, 60.
53 Sukanta 2014, p. 81.
54 Toer 1982.
55 Hersri Setiawan, interview, Jakarta, 27 April 2007.
56 Siwirini 2010, pp. 96–98.
57 Winata (pseudonym), interview, East Java, 22 February 2008.
58 Vickers 2010, p. 57.
59 McGregor 2012, p. 245.
60 Sukanta 2014, p. 100.
61 Tempo 2013, p. xx.
62 Farram 2014, pp. 1–24.
63 Dwyer 2011, p. 230.
64 Alit et al. 2015, p. 56.
65 Hendy, *Noise*, 384.
66 Sukanta, *Breaking the Silence*, 335.

References

Alit, A. et al., 2015. *Prison songs, nyanyian yang dibungkam*, Denpasar: Taman 65.
Amnesty, 1977. *Indonesia: An Amnesty International report*, London: Amnesty International.
Classen, C., 1993. *Worlds of sense: Exploring the senses in history and across cultures*, London and New York: Routledge.
Corbin, A., 1995. *Time, desire and horror: Towards a history of the senses*, trans. J. Birrell, Cambridge, UK: Polity Press.
Corbin, A., 1998. *Village bells: Sound and meaning in the 19th-century French countryside*, trans. M. Thom, New York: Columbia University Press.
Cribb, R., 2001. 'Genocide in Indonesia, 1965–1966', *Journal of Genocide Research*, 3, no. 2, June, pp. 219–39.
Crouch, H., 1988. *The army and politics in Indonesia*, revised edn, Ithaca, NY: Cornell University Press.
Dwyer, L., 2011. 'Building a monument: Intimate politics of "reconciliation" in post-1965 Bali' in A. L. Hinton, ed., *Transitional justice: Global mechanisms and*

local realities after genocide and mass violence, Piscataway, NJ: Rutgers University Press, pp. 227–48.

Farid, H., 2005. 'Indonesia's original sin: Mass killings and capitalist expansion, 1965–66', *Inter-Asia Cultural Studies*, 6, no. 1, pp. 3–16.

Farram, S., 2014. 'Ganyang! Indonesian popular songs from the confrontation era, 1963–1966', *Bijdragen tot de Taal-, Land- en Volkenkunde*, 170, no. 1, pp. 1–24.

Harsutejo, 2013. *Keluarga abangan: Memoar*, Bandung: Ultimus.

Hatley, B., 2008. *Javanese performances on an Indonesian stage: Contesting culture, embracing change*, Singapore: NUS Press.

Hearman, V., 2009. 'The uses of memoirs and oral history works in researching the 1965–66 political violence in Indonesia', *International Journal of Asia-Pacific Studies*, 5, no. 2, pp. 21–42.

Hearman, V., 2012. ' "Dismantling the fortress": East Java and the transition to Suharto's New Order regime (1965–68)', PhD Thesis, The University of Melbourne.

Hendy, D., 2013. *Noise: A human history of sound and listening*, London: Profile Books.

Heryanto, A., 2014. *Identity and pleasure: The politics of Indonesian screen culture*, Singapore: NUS Press.

Kutzler, E. A., 2014. 'Captive audiences: Sound, silence, and listening in Civil War prisons', *Journal of Social History*, 48, no. 2, pp. 239–63.

McGregor, K., 2012. 'Mass graves and memories of the 1965 killings', in D. Kammen and K. McGregor, eds., *The contours of mass violence in Indonesia 1965–68*, Singapore: NUS Press in association with the Asian Studies Association of Australia, pp. 234–62.

Moeljono, D. S., 2013. *Banten seabad setelah Multatuli: Catatan seorang tapol 12 tahun dalam tahanan, kerja rodi dan pembuangan*, Bandung: Ultimus.

Prior, J., 2011. 'The silent scream of a silenced history: Part one: The Maumere Massacre of 1966', *Exchange*, 40, pp. 117–43.

Rochijat, P., 1985. 'Am I PKI or non-PKI?', trans. B. Anderson, *Indonesia*, 40, pp. 37–56.

Roosa, J., 2006. *Pretext for mass murder: The September 30th movement and Suharto's coup d'état in Indonesia*, Madison, WI: University of Wisconsin Press.

Roosa, J., 2008. 'The truths of torture: Victims' memories and state histories in Indonesia', *Indonesia*, 85, pp. 31–49.

Setiawan, H., 2006. *Kidung untuk korban: Dari tutur sepuluh narasumber eks-tapol Sala*, Sala: Pakorba Sala.

Siwirini, S., 2010. *Plantungan: Pembuangan tapol perempuan*, Yogyakarta: PUSdEP, Sanata Dharma University.

Smith, M. M., 2007. *Sensory history*, Oxford and New York: Berg.

Sukanta, P. O., ed., 2014. *Breaking the silence: Survivors speak about 1965–66 violence in Indonesia*, Melbourne: Monash University Publishing.

Tan, S. L., 2010. *G30S 1965, Perang Dingin & kehancuran nasionalisme: Pemikiran Cina jelata korban Orba*, Depok: Komunitas Bambu and Lembaga Kajian Sinergi Indonesia.

Tempo, 2013. *Pengakuan algojo 1965: Investigasi Tempo perihal pembantaian 1965*, Jakarta: Tempo Publishers.

Toer, P. A., 1982. *This earth of mankind: A novel*, Melbourne: Penguin.

Vickers, A., 2010. 'Where are the bodies? A transnational examination of state violence and its consequences', *The Public Historian*, 2, 1, pp. 45–58.

Weintraub, A. N., 2010. *Dangdut stories: A social and musical history of Indonesia's most popular music*, New York: Oxford University Press.

Interviews and Personal Communication

Harsutejo, interview, Jakarta, 30 April 2007.
Hersri Setiawan, interview, Jakarta, 27 April 2007.
Ling (pseudonym), pers. comm., 16 February 2009.
Marcus (pseudonym), interview, Jakarta, 3 February 2009.
Muhammad (pseudonym), interview, Bangil, 12 February 2009.
Oei Hiem Hwie, interview, Surabaya, 2 August 2007.
Winata (pseudonym), interview, East Java, 22 February 2008.

Films and Audio-Visual Material

The act of killing, 2013. Directed by Joshua Oppenheimer; produced by Signe Byrge Sorensen. Copenhagen: Final Cut for Real.
Jembatan Bacem: Film dokumenter tentang peristiwa 1965, 2013. Directed by Yayan Wiludiharto. Surakarta: Perkumpulan ELSAM and Pakorba.

10 "For a Few Seconds, Imagine"

An Aural Experience of Six Days
of Terror at the Stadium of Chile,
12–17 September 1973

Peter Read

In the most recent signage at the National Stadium, Santiago, Chile, visitors are invited to experience what the blindfolded detainees endured by imagining their sensory world:

> *For a few seconds, imagine raucous military marches filling the air in a vain attempt to disguise the sharp bark of rifles and the deadly thudding of the machine-guns and the terrified cries of the victims. . . .*
>
> *The Stadium, barracks and prison, where artillery points at the entrances of the sporting complex. The [military] patrols tramp on the pavement of the interior paths allowing the echo of orders, the sounds of greased metal, shouts and discharge of rifles and heavy machine-guns. The gunpowder stinging in the nostrils paralyses the prisoners' hearts in the cells like Gate 8. The three long-haired teens were taken away with their heads covered under a blanket and they never returned to be healed at the [exit] gate. Neither did the two workers in Dressing-room 4 taken off to interrogation in the frightful Velodrome. Many years later, we understood that their relatives found their mortal remains.*
>
> (Signage, National Stadium of Chile,
> Santiago, installed March 2014)[1]

Hearing, not sight, was the principal sense by which the detainees of the dictatorship in Chile that seized power in 1973 often made sense of their surroundings.[2] Almost all detainees were blindfolded shortly after their arrest, when transferred from one detention centre to another and when they were being interrogated and tortured; and some remained sightless for the whole of their time in custody. Blindfolded men and women held in the House of Memory, Londres 38, in the city's central business district, heard the church bells of the St Francisco church outside and realized that they were imprisoned in what had been, only weeks before, the headquarters of the Socialist Party. Roberto Sanchez, a detainee at the National Stadium, but a worker at the same place only a few days before, calculated by the distance to where he was frogmarched that he must be at the Velodrome, and then divined, by the cries of the women 50 metres away, that they were being tortured and interrogated in an adjacent building to his own. Alone among the prisoners that day, he knew exactly which building it was, and can identify it still. Detainees in the infamous tower at the Site of Conscience, Villa Grimaldi, heard the children of the guards splashing in the pool. The detainee Gladys Diaz survived unspeakable tortures to recall 'the ways that one finds to defend oneself are unlimited. I sometimes dreamed about beautiful things. . . . I remember having awakened to the sound of a little bird that was outside, and how I was able to keep the sound of that bird's singing in my ears for days.' The Security Services used the sound of gunfire as a weapon of terror. Several blindfolded detainees recall hearing what they suspected to have been fake executions by automatic weapons at the Velodrome, while troops arriving at another clandestine torture and execution centre, 'Nido 20,' fired their guns on arrival to keep civilians away from the windows of their homes.[3]

In this chapter I take up their challenge to 'For a few seconds, imagine . . . ,' to reproduce the aural ambience of another Santiago Stadium, known at the time of Pinochet's coup as 'The Stadium of Chile.'[4] I propose to follow, through metaphorical headphones, the sounds in this Santiago Stadium, where students and workers were held for six days in the very first week of the coup. Between 12 and 17 September 1973, this edifice imprisoned more than 5000 detainees, at least 100 of whom were interrogated, tortured and executed within the walls. The rest were then escorted to the much larger detention centre at the National Stadium.

Pinochet's coup followed three years of progressively greater disturbance, which began with the election of the Socialist President Salvador Allende in November 1970. On his election, many left wing parties in Chile held out enormous hopes for root-and-branch reform, including the confiscation of large feudal estates and nationalization of the country's copper mines. Stringent opposition from the right soon stalled Allende's ambitious programs. Inflation soared while traditional allies staged strikes against their government. Allende seemed to stumble from one crisis to another. Right wing opposition forces predicted complete national anarchy, to be shortly followed by the imposition of a Cuban-style 'people's republic.' Most senior officers of the armed forces shared the same fears and a full six months before the coup, began to examine the possibility of a right wing overthrow of Allende's elected government. The carefully planned coup took place at dawn on 11 February 1973.

Why imagine the sounds only in the Stadium of Chile, the sole detention centre in which blindfolds were *not* used? One reason is that the focus on the terrible fate of the internationally famed singer-songwriter Victor Jara here has tended to distract from the experiences of the thousands of others detained with him at the National Stadium. Apart from Jara, we know much less about the events of the Stadium of Chile than any other major centre of the dictatorship. For this reason I have not 'overheard' much of the sound concerning Jara, most of which, in any case, took place downstairs, out of the range of our imaginary microphones.

Equally important, while hundreds of individuals have left oral testimonies of other detention centres, in the Stadium of Chile their experiences were different. In this confined space everyone witnessed, and many later narrated, exactly the same events, perpetrated within a single small space and within a very short time. Their different viewpoints we can take to be, quite literally, those that were provided by whatever part of the stadium in which they were forced to sit; so their experiences and memories were almost the same. Many of these have been recorded, transcribed and published, and I have used them literally: simply to imagine the exchanges between the military and the detainees would be deeply insulting to the survivors whose memories we now have. Every individual vignette that follows—the murder of the detainee mishearing an order, the informer who lost his reason, the weeping conscript perhaps with only months of training, the rantings of 'the Prince' (Edwin Dimter), the virtual loss of control by the troops and the long narrative of the prisoner taken to the infirmary—all are recorded in the oral or written histories. I have reproduced their memories—though as whispered conversations with other detainees—as they recalled them.[5] Their published words give them the agency of which they were so brutally deprived and provide the backbone of this chapter.[6]

What follows, though, is a brutal piece, almost devoid of friendship or compassion. I have done this deliberately, for it is these elements of the incarceration that the survivors most vividly recall: the terror of the unknown, the bewilderment of the conscripts; and the raw violence, tension, brutality, irrationality and hatred that the regular soldiers wrought collectively on the detainees. They are the individual recollections, in sound as well as sight, of solitary men and women confronted with violence directed at each of them personally as well as at the group. In this profound sense, each sensory experience was solitary. The memories of those who actually *were*, in other detention centres, blindfolded, rehearses that same sense of total powerlessness, a world of sinister, violent or menacing sound in a nightmare of an irrational and incomprehensible unknown future.

Direct translations from vernacular Chilean Spanish are frequently impossible, especially the language of abuse. Spanish is rich in sexual insults that carry distinct cultural overtones. 'Maricon,' which was shouted by the detainees in the stadium as they watched a soldier savagely beating one of their number with a rifle butt, means at one level 'poofter,' but it implies also that he (never 'she') has betrayed his sex and is therefore also a coward and a traitor as well as a homosexual. Epithets like 'sewer scum' can be almost comical in English, while 'son of a whore' (never 'daughter of a whore'), may not mean much in Australian English. In Latin America the phrase is a deadly insult, traducing not only the honour of a particular mother, but also the sanctity of all women. English, rather poor in really deadly invective, cannot go much beyond 'cunt' or 'motherfucker,' insults which already have their equivalents in Spanish. I have therefore sometimes used Australian equivalent swearwords and vernacular rather than translating the Spanish directly, conscious though I am of their inadequacy.

At the moment of the coup, 11 September 1973, Augusto Pinochet's immediate intentions were threefold: firstly to unseat the government by eliminating President Allende and disposing of his senior ministers and officials; secondly to raid the radical working class settlements known as *poblaciones* in order to detain known members of political parties supporting Allende, especially the Communists, the Socialists and the armed revolutionary party known as the MIR; thirdly he needed to discover the whereabouts of other leading members of these parties, and the location of their alleged arms stashes. The Pinochet plan demanded simultaneous action on many fronts: sealing the radical enclaves, capturing the wanted, transporting them immediately to a large holding centre and initiating interrogations, torture or summary execution.

The major short-term destination for the detainees about to be apprehended on 11 September would be the Stadium of Chile. The building stands in a run-down part of the city, near some of the most notorious *poblaciones*. Conveniently, it also stands six blocks from a radical university, the State University of Technology, regarded as subversive and dangerous. The University is known by its acronym 'la UTE,' and compared by its critics to the Paris Polytechnic.

Pinochet planned daybreak raids on both the *poblaciones* and the UTE.

On the very day of the coup, Allende was scheduled to appear at the UTE to declare a national plebiscite to revalidate, he hoped, his position as President of the Republic. His Master of Ceremonies would be the internationally famous entertainer Victor Jara, loved by the left, but detested by the extreme right. Jara would not be alone. In anticipation of the plebiscite announcement, thousands of Allende supporters assembled on the campus; then, as the news of the coup spread through the nation, they remained at the university to oppose it as best they could. At mid-morning on 11 September, the Vice Chancellor advised staff and students to leave the campus immediately, to get to their homes, their prepared safe houses or their party headquarters. By nightfall the hundreds who had chosen to remain would be trapped by the security forces encircling the campus and next day would be marched or trucked six blocks to the Stadium of Chile.

In 1973 the Stadium was a popular venue for smaller-scale concerts, boxing tournaments, ping-pong competitions and basketball. No spectator was too far away from the action. Very many of the detainees about to be held there had happy memories of sports events and concerts. Victor Jara himself had performed there recently amidst much music and gaiety. But on 12 September, having spent a night of terror on the campus, the staff and students were stumbling into the stadium that now sounded much more menacing. They include the Vice Chancellor and Victor Jara, who was still, as yet, unrecognized.

Miking up the Stadium

The stadium seating is arranged as two-tiered banks of seats opposite each other, and two smaller galleries also facing each other, all looking down on the central arena. Thus, the configuration is more that of a boxing ring than a concert hall.

To catch the full ambience of the public spaces we install microphones in the four upper corners of the building, just above the topmost row of seating, set to an optimum response of 90 degrees. As channel five, we swing an omnidirectional microphone from the centre point of the ceiling, to hang three metres above the arena. We need wall pick-ups, channels six and seven, for both the two long entrance and exit corridors: the right-hand customarily used for entry from the foyer; the left, from which the changing rooms and ablutions also are accessed, for exit. To complete our eight-channel feed, we place an ambient microphone in the entry foyer. What is not miked, though, are the two floors of subterranean changing rooms, the refectory and dormitory downstairs. These are out of sight and out of microphone range. That area is forbidden even to police and conscripted soldiers. Only selected members of the security forces may enter. No cameras, no microphones. No records.

Sound check. Even when empty, the stadium produces a singular acoustic. The whole precinct is enclosed by a high-domed ceiling from which interior sounds bounce and resonate distantly off the hard surfaces. Because of the thin walls of the stadium, exterior rumbles of passing vehicles, even street sounds, can disturb the vibrations within. When the traffic is still, one feels as well as hears the close atmospheric ambience, like that of a temple. The air seems almost to tremble from minute but constant reverberations. This is a silence that is tactile as well as audible.[7]

From 12 September to 17 September 1973, the stadium will be anything but silent. We are about to audition the six most terrifying days in the 17-year history of the Pinochet dictatorship.

Wednesday, 12 September 1973, 3 pm

The greatest commotion is coming through channel 8 (entrance foyer). An extraordinary number of people seem to be crowding into it. Amidst the maelstrom it is hardly possible even to distinguish the military orders, but it's clear from the increasing volume from channel 7 (wall pick-up, right corridor) that hundreds of detainees are being herded from the foyer into the stadium itself. Through both channels comes a clatter of steel-studded army boots on tiles, mixed with softer shoes in a huge crush of humanity. Driving both mikes into distort, the detainees are communicating their terror in sound: cries, shrieks, oaths, heavy blows. An officer's voice:

Drag him out of the way.

Heavier blows, moans of fear, automatic weapon fire.

Line up here.

The pandemonium grows in intensity. The panicked cries and the most aggressive military voices seem to be focused on the bottleneck, but by fading down the foyer mike and bringing up channel 7, the narrower space and shorter echoes begin to isolate individual voices. Someone slips—on what?—and falls heavily with a cry.[8] A blow. Two more blows, a groan. Another volley of an automatic weapon reverberates round the amphitheatre and must be clearly audible outside. Obscenities. Crying, shouts of terror, alarm, panic, abuse.

> *Cigarettes and lighters in that pile.*
> *Hands behind your necks.*
> *Señor . . .*
> *What's your name? Get over there!*
> *Please sir . . .*
> *Get over there.*
> *Hey you. Stop. Come back here.*
> *Wait a min . . .*
> *Yes you.*
> *Can I just . . .*
> *In there you swine!*

It's clear there's something truly terrible happening in the right entrance corridor, a thud of bodies falling, or tripping, or slipping or knocking against each other. Grunts, curses, wails, male and female. A multitude of human footsteps, some running, stumbling, staggering, no regularity even of military boots. In the arena a confused shouting coming through on the arena mike, channel five, until three single shots reverberate—Fired where? At whom? They momentarily still the tumult. An officer bellowing:

> *UTE scum into that green corner. Move it. Move it. Fill up the back first.*
> *Keep still. First one to move is killed. First one to talk is killed.*

The volume unit metre on channel 1, upper gallery left (east) is flickering. Shuffling feet indicate that the seats are being taken up from the top. The dull thunk of heavy greased metal parts being fitted together comes loud and very close to the channel 1 position. A well-educated military voice:

> *Get it level to train on every one of these bastards. And line up on that*
> *fucker with the red shirt.*
> *Yes sir.*

Someone sitting very close is weeping. A whisper, very soft. It sounds like:

> *Compañero, tranquilízate. Keep calm my brother.*

Sobbing and cries continue from many locations. Channel 2, upper gallery north, is also now picking up many seats being occupied.

The uproar from channel 6 in the corridor below is louder than ever, but now channel 7, outside the left-hand corridor, is beginning to suggest a more ordered universe. Military orders, detachments of men—some of them raw conscripts by the sound of their un-coordinated movements—carrying equipment. Heavy objects scraping the tiles are being dragged about. The harsh grinding of several heavy steel drill-bits on concrete booms hollowly into the auditorium and is picked up on all channels. Soldiers must be drilling holes in the walls of that left-hand corridor. Or can it be the floor?[9] Orders from the middle distance:

> *Yes in that corner. Not too close.*
> *Hey you. Out. No police or conscripts in here. Named security only.*

Only military personnel here, at least 20 or 30, but the purpose of their sinister preparations is unclear.

Same Day, Two Hours Later

Loud through channel 5, above the arena, hundreds of feet are running, shuffling, limping. More detainees are occupying the seating on the right; that section sounds almost full. A buzz of subdued whispers underlies the military shouting. An amplified command that must be coming from a megaphone, impossible to tell its location:

> *Move it, move it. UTE on the left, worker scum on the right.*
> *Por ayá huevones. Get over there you motherfuckers.*

Another burst of automatic fire is deafening on channel 5. Sobbing. More distant yells, obscenities. The distorted Stadium PA crackles and booms through every channel:

> *No noise. No moving. Anyone who moves will be killed on the spot.*
> *Now look up on every corner. That's a .45mm machine-gun trained on every one of you. The experts call this machine-gun 'Hitler's saw.' It can open up a man and cut him in half.*
> *Fellers, let's make a deal. Nobody move.*

The whole stadium falls suddenly silent.

> *On the other hand, we want to use it.*

The same voice screams louder than before:

> *You are the vilest scum in Chile. The sewer of the nation. Don't expect any pity from us.*

Thursday, 2 am

Fade up channels 1 and 2, the seating tiers holding the greatest number. UTE students and staff are mainly on 1, though it's no longer possible to distinguish them from the mostly inaudible subdued whispers from the factory workers and *poblaciones* on 2. Bring up channels 3 and 4 on the opposite side. On full gain we can hear higher pitched voices. Was that a baby's cry? There are mothers there! There must be a space reserved for women detainees in the middle. Who's on the right? Louder subdued whispers now, but in accents that are not Chilean. They are Uruguayan, Cuban, Argentinian. Mexican. One sounds as though he may be an American. Bring up channel 4. A separate group. They must be the ones thought to be most dangerous, already tortured, scheduled for further interrogation.[10] Someone's whispering near the mike:

> *Jesus, it's so cold. Compañero, you've got a blanket. What about sharing it. We've got to cover ourselves in something to conserve a bit of energy.*

The public address system:

> *Marxist shit. On your feet, traitors. Fernando Flores. Jorge Godoy Godoy, Mario Céspedes Gutiérrez.[11] Martínez. Down here on the double.*

The Same Day, 9.30 pm

The full-bodied atmosphere indicates that the stadium is nearly full now, even the arena often doubling as a ping-pong space itself sounds crowded. But the ambience of this throng of 6000 people is not that of a normal crowd. From all the galleries only crying or whispered conversations too hard to pick up. The only clear voices are raucous and military.
    The loudspeaker crackles. An amplified voice.

> *Listen you Marxist sons of bitches. Can you hear me? On your feet!*

Creaking seats and groans.

> *When you hear me say, 'Can you hear me?' You will answer 'Si si señor we hear you.' Now. Can you hear me?*

A ragged assent.

> *This is the voice of the Prince. I am the Prince.*
> *Can you hear me?*
> *Yes sir, We hear you*
> *Can you hear me? . . .*

Cutting through another assent, through channels 1 and 2, the sounds of a scuffle, shouts, running feet, a long high-pitched scream.

Silence, before a huge collective moan of horror rises and falls almost uniformly from the entire stadium.

> *Get a stretcher. Take it away.*
> *Halt Halt!!*

A shot.

A stentorian order loudest through channel 5, arena:

> *Get down here. Yes, you.*

The far right-hand microphone, channel 2, is picking up the factory workers' breathing, shuffling and whispering:

> *Jesus, see who that is?*
> *It's some taitor. Picking us out. Identifying us. God almighty. Keep your*
>   *head down.*
> *You! Yes you!! Get down there.*

Military feet are approaching. A seat is vacated from somewhere maybe two metres in front of the microphone. Whispers.

> *Señor ayudale. God help him.*
> *Compañero. Be strong.*

The footsteps of the patrol are loudest on channel 7, left corridor, then fade, but the whole auditorium ambience is louder, no longer so submissive or terror stricken. An audible hostility is rising on both sides.

Thursday, 13 September, 8.30 am

An anguished voice screams from a high gallery, from the same side from where a detainee launched himself yesterday.

> *The people have fallen. Here come the poofter-traitors. Long live Chile.*

Another piercing howl, and a second later a thud from the arena mike channel 5, as a human body lands on tiles. A scream of pain, two, many; screams like no other, higher and higher in pitch. A cry of grief from the upper level on the right. Swearing from the corridor emptying into the arena. Blows, grunts, savage blows, cries, two pistol shots.

Through the loudspeakers:

> *Nobody move!!*

Bringing channel 1 up to max and closing the others we're picking up snatches of a conversation that's not meant to be heard. It's the voice of the informer:

> *. . . the one you want . . .*
> *The feller with the beard?*
> *. . . one sitting next to him as well. Both bad bastards.*
> *. . . anyone else?*
> *Yeah, see that . . .*

Channel 3 is picking up the sounds of a scuffle. A loud but unamplified military voice.

> *That's him. Yeah. Next to the gringo. Bring him here. Now.*

Simultaneously, through the right corridor, channel 6, the voice of the officer who calls himself the Prince, shouting so loudly he's coming through on every channel except the foyer:

> *Right you Cuban pig I'm going to take care of you personally.*

A cry of agony.

> *See this ear? You won't need that any more. Now. Into your mouth it goes. Now go back and tell Castro this is how we treat Marxist sewer rats in Chile. If you ever do get back.*

Screams and more screams as he's dragged off, loudest on channel 7 left. Savage kicking, boot on bone, heavy metal on flesh. An order no longer an order but a scream.

> *Get him downstairs!*

The screams grow fainter. A metal door slams in the middle distance. The crowd is hushed, knowing what to expect. A full 15 seconds before a short and indistinct burst from an automatic weapon. In the main auditorium, indistinct murmurs and weeping, brutally interrupted by another grunting floor-level struggle, this time from near the arena mike. The suicide has not managed to kill himself, but is still crawling around on the floor. Metallic clacks as weapons are cocked.

> *Come and find me you mothers' cunts!*

A shuffle of military feet. The prisoner's insults grow louder as we hear him being dragged into the left corridor, then dimmer. Followed by another expectant silence in the stadium. No shots though, only screams.

Same Day, 11.30 pm

There must be so many people crammed inside that the arena is no longer a thoroughfare, but packed with detainees. Whispers are easiest to pick up through the channel 5 mike suspended above them because it must be much harder there for the guards to see individuals.

> *I can't get to sleep because of those fucking spotlights.*
> *Try putting your hands over your eyes.*
> *No use.*

Nobody can sleep for another reason, that the loudspeakers never stop bellowing orders:

> *Gregorio Mimica Argote. Identify yourself.*

Mimica Argote must be standing up. The sound of four military feet in a hurry climbing the stairs.

The footsteps fade through channel 7 but no metal door slams, only screams rising above other screams of terror and agony like few in the whole stadium have ever heard before.

In the arena:

> *If you get called up you don't come back.*
> *Is that what happens when they call you?*
> *Yeah.*
> *They beat you to death?*
> *No, they get rid of you during the interrogation.*
> *What do they do with the bodies?*
> *They're saying they just chuck them in the street. Or in the river.*

A dull volley, loudest from the left corridor, echoes off the roof, so many echoes it's hard to determine where it's coming from.

> *That'll be another compañero now.*

A prolonged whisper from someone who seems to be a new arrival:

> *They picked me up first and took me to the Seventh Commisariat and threw me onto this electric mattress asking me all these dumb questions. Some kind of machine that they had to get the power through by turning this crank handle. No idea what it was.*
> *Shush. . . . Wait . . .*
> *But it wasn't working properly and they lost it and got mad and they started to try and find out what was wrong with it. So in the end they brought me here. Four days with nothing to eat. And whatever*

there was did no good for my ulcer. I couldn't even sleep because of the pain. So someone must have recognized me and said to take me to the infirmary. He said to a conscript—'Take him there and bring him back'—emphasizing this—'don't leave him down there,' because most of them who go there don't come back.

Hang on . . .

So we went down this staircase. Looked like a long passage filled with military of one sort or another. And another line of compañeros just as long all sitting down with their backs against the wall. Every one had a soldier in front in charge of them making them stay there. Must have been about four hundred people there, all up. Then we come to this sort of hall.

The screaming of tortured bodies, loudest on channel 7, is now so common that the speaker continues without interruption:

Three big piles of bodies, four deep, lined up side by side, then four crossed over on top of them and four on top of them. I reckon about between 32 to 40 in each pile. I thought at first they must have been some kind of mass torture, later I realized they weren't moving or breathing.

The screams from the ablutions rise higher in intensity and pitch, until they are barely recognizable as human. They can only come from a person in extremis undergoing electrical torture, the electrodes connected to the tongue, the toes, the genitals.

God let them rot in hell forever.

I get to the sick bay and find this friend from my district and I ask him a dumb question, 'Are they going to interrogate these compañeros? Why are they here?' 'No, they're corpses. Killed during interrogation.' I ask him what they were going to do with them and he says he'd heard an officer say they were going to chuck them in the street or in the river. Then they gave me an injection and I came back up again.

Hey, watch out.

Don't piss any these cunts off or we'll all be down there.

Simultaneously, a whisper coming through from the wall mike, channel 6, in the right-hand corridor, a conversation from the male toilet:

Christ what a stink. You can see they're not too worried about comfort. The taps are all stuck over and water's going everywhere.

Barking orders from several metres away. The same voices much closer, and softer:

> *You're Victor Jara aren't you? Dios mio, what have they done to you?*
> *Yes, compañero.*
> *How are you making out? I think they know who we are round here.*
> *How did they treat you like that?*
> *Very bad. . . . (inaudible whispers)*
> *We're totally fucked.*
> *More or less (inaudible)*
> *But listen, they say there's a spy here. Be careful what you say, he could be anybody.*

Authoritarian footsteps are approaching and stop. An officer's voice:

> *What are you two pigs talking about? Hey, you're that fucking singer, aren't you? Victor Fucking Jara.*

Shuffling feet, groans, execrations.

> *Detail!!*

More troops enter the toilet. Amid the confused uproar of shouting and scuffling, a voice clearer than the rest: it's the Prince again.

> *What's this bastard doing here? Don't let him move from here. He's reserved for me.*

A series of blows, metal on flesh, boot on bone.

> *All right, get him out of here. In there.*

Corridor wall-mike, left, channel 7 carries yet again the sinister and fading reverberation of cement and tile and steel door.

Friday, 14 November, 4 am, Main Arena

Arena microphone channel 5.

Jara evidently has been returned from the subterranean changing rooms, in bad shape.

> *Compañero Victor, tranquilízate. Let me wash your face and get some of this blood off now. Oh, sorry hermano [brother].*

A strained and painful gasp, barely audible:

> *Paper . . .*
> *Paper. Right. I'll get some. Who's got a pencil?*[12]

Three minutes later. Channel 5, arena microphone:

> *I want to go to the toilet sir.*

The uncertain voice of a conscript:

> *OK, I'll get the officer.*

Clumping, authoritarian feet.

> *On your feet. Hands on your neck. Get going.*

Shouting to everyone:

> *Everyone keep still.*

But we can hear the hesitant footsteps of the detainee still walking at the same pace. Has he not heard the order?

> *I told you to keep still.*

Two shots. A body falls. A collective sigh of anger, pity, horror and terror.
Instantly, as if in answer, running feet from channel 2, upper gallery, from where the workers are seated. A scuffle. More running feet from many directions. Three heavy blows of metal hitting flesh. Another collective sigh on every channel, but more of anger than terror. A loud voice through upper gallery, channel 4, from where the 'dangerous' prisoners are being held.

> *You cowardly cunt.*

Other voices, from every point:

> *Poofter!*
> *Bastard!*
> *Motherfucker!*
> *Traitor!*

The cry echoes round from all channels. A whistle. Again the military are on the point of losing control. Running military feet, then they stop dead. The sound of many guns being cocked.

A detainee through channel 1, shouting:

Get down everyone!

A volley of shots from the arena: who knows at whom? One of 'Hitler's saws' opens up with a sustained burst a metre from the channel 2 mike, blowing the meter off the dial. The whole stadium is edging into anarchy. A mutinous, angry silence. Quiet from the entrance corridor and the ablution sections. The foyer mike, channel 8, picks up, maybe a kilometre away towards the naval station at Quinta Normal, a prolonged burst of automatic fire. Answering shots, a little closer, sirens. A light truck roars by. Then another: they're the small Chevrolets the military use for detainees and to get their troops around quickly.

Thursday, 13 September, 1.30 pm

In the main arena.

Where's that bastard Jara gone?

The voice of the Prince again:

We've got a place for you. Upstairs.

Up the stairs, higher, between channels 1 and 2 in the upper seating. They must be heading for the broadcast box.

Let's bust the hands of this mother's cunt.

The sound of wooden stick hitting human flesh, over and over.

Sing now, you bastard. Get up.
Bend his wrists over that bar.
Want a cigarette, hijo de puta [son of a whore]? I'll light one for you.
  Here you are . . .

A long shudder of agony.
   Channel 1, about five metres from the mike, is picking up whatever's happening in the broadcast box. Everyone seated nearby also can hear the screams.

They're taking Victor apart.
God they are too.
Compañero adios.
Stay strong.

The Main Arena, 3.30 pm

After the beating two hours ago, it's incredible that Jara is still alive, but he's croaking out some kind of song, only snatches audible:

> *One dead, with a blow like I never believed*
> *Could be dealt to a human being . . .*
> *Oh my God, is this the world you created*
> *Was it . . .*

Friday Morning, 11 am, Loudspeakers

> *Marxist swine. This is the Prince. Do you hear me?*
> *Yes sir. We hear you.*
> *The Armed Forces only want to punish those involved in the former government. The great majority of you were in the Popular United Party only out of idealism. You have nothing to fear. To facilitate matters it's necessary to point out the dangerous ones. These are the ones on our list so far. So tell us who they are. Come out here. Littré Quiroga . . .*[13]

Silence. Nobody's coming out.

> *Go and get him then.*

Tramping feet from all directions. There must be at least a dozen pairs of boots coming towards the detainees, up all the upper gallery channels.

> *Get them downstairs.*

The captures are interrupted by another disturbance coming through on channel 2, upper western gallery. Shouting, a scuffle. Running feet. At least two people, grappling with each other.

> *It's someone wrestling with one of the guards.*

Shouting coming from all around, and military boots running.

> *It's that bloody traitor informer again. He's trying to grab his gun.*
> *Mad as a cut snake—*
> *Give it back. Give it . . .*

No regular soldier would plead like that. It's another young voice, Villarrica accent from the far south, of someone young and uncertain.
　　A shot.

Staggering. Someone's gone over the balcony. But who? Or is it both?
Another sickening thud of another body on concrete, brutal on channel 5, left-hand corridor, just below. Voices . . .

Stone dead.
Crazy.
Somehow he thought he was a guard.

Up above in the gallery, it's suddenly quiet. Except for the sound of someone sobbing uncontrollably.

Ay mi madre. Ay mi madre.[14]

It's the voice of the conscript.

Saturday, 16 September, 11 pm

One by one the upper channels are growing silent, except for the stadium's mysterious audible ambience. The detainees are leaving, through the same corridor they entered five days ago but in a fashion much more orderly than their arrival. No screams, fewer protests, only military orders. The detainees seem sunk in a sullen silence.

Sunday, 17 September, 3 am

Channel 8, foyer, is picking up much activity: trucks, orders, reversing, brakes, revving, troops. The trucks are much closer to the foyer than the trucks and buses taking the detainees this morning. They are heavier vehicles too; they sound more like heavy transports.

Together. . . . Up.

The troops are throwing heavy objects up on the trucks.
Five-thirty am and the last truck is leaving.

Same place sir?
Yes, same place. Right down the back. You know where to go.[15]

By 6 am, all channels are silent except the left corridor, channel 7, which still carries the sounds of subterranean military violence in the middle and far distance. Gunfire. Running feet. Marching, distant orders. Long howls of torment, but far away, and getting fainter and with odd echoes that bounce weirdly but faintly round the empty auditorium. A heavy steel door slams somewhere downstairs. Gunfire, in single shots. Then more, many more.[16]

An unforgettable element in Arthur Koestler's *Dialogue with Death* is his account of prisoners captured during the Spanish Civil War waiting day after day in their cells. They have not the slightest idea if today they will be set free, allowed to walk in the exercise yard, transferred to an undisclosed destination or executed.[17] They do not know why they are imprisoned and have only a vague idea of by whom. They do not know the charges—if any—under which they are held, nor who—if anyone—has made them. Even after summary trial they have no idea of their fate.

Detainees held blindfolded, according to their own testimonies, suffer from the same sense of what we might call 'present-time deprivation.'

Forty years later, at Santiago's National Stadium, detainees were blindfolded, handcuffed, frogmarched 200 metres from the changing rooms in which they were being held and forced to stand at attention for hours in the sun listening to the screams of the tortured, before their own interrogation and torture began. The survivors wrote, in 2014:

> For a few seconds, imagine raucous military marches filling the air in a vain attempt to disguise the sharp bark of rifles and the deadly thudding of the machine-guns and the terrified cries of the victims. . . . The [military] patrols tramp on the pavement of the interior paths allowing the echo of orders, the sounds of greased metal, shouts and discharge of rifles and heavy machine-guns.

If Koestler's gift to the world in the first half of the twentieth century was to reify the terror of the prisoners unknown, then the Chilean testimonies of the blindfolded reify the same terror in the second half of the century. The detainees I describe in this chapter were not blindfolded, but they may well have been, for they suffered the same terror of not knowing what was to happen to them next week or tomorrow, or in five minutes. My chapter tries to re-create what is perhaps this most elemental nightmare of detainees held by rogue states in our time, the terror of an unknown future. For a few seconds, imagine.

Acknowledgements

I thank the organizers of the conference on 'Sound in History,' Joy Damousi and Paula Hamilton, who made this article possible. Juan Medina and Nena González consented to guided tours and interviews. As ever, Marivic Wyndham contributed invaluable historical knowledge and cultural understanding. Thank you.

Notes

1 For discussion on the significance of the change of style in the memorialization of Chile's sites of Consience, see Read and Wyndham 2016
2 In Latin America, 'detainees' rather than 'prisoners' is used customarily to distinguish those illegally detained by the state from ordinary criminals (Other authors or editors whose work I have relied on are Anon, Cuevas 2009, Fuentes, Sepúlveda and San Francisco 2010, Jara 2001, Navia nd, Peña 2008, Quiroga Cavajal 2010, Ruz de las Paños 2007, Villegas 2013, Zuñiga 2008).

3 The material in this paragraph is drawn from visits to the National Stadium of Chile in 2006 and 2014 and Nido #20 (2013), testimony and interviews with Michelle Drouilly and Roberto Sanchez and transcribed interviews with Gladys Díaz, in Camiragua; and Luz 1993; see also Aránguiz 2006.

4 The Stadium of Chile (now known as the Victor Jara Stadium) was the largest detention centre in Santiago in use immediately after the coup. One week later, detainees were moved to the internationally infamous detention centre, the National Stadium. This paper refers to events in the first of these holding centres, the Stadium of Chile.

5 The interviews include those drawn from Villegas 2013, in which are transcribed the interviews with 'C. M.,' Esteban Carvajal, Julio Peña, 'Milico' Solano, Yusufi, César, Patricio, Vladimir Tapia, J.M. Wilson, A. Monsalve and C. Orellana. The survivor Juan Cristóbal Peña carries out a brief guided tour of the stadium in the documentary *La Funa de Victor Jara*. I also draw upon impromptu tours conducted by an unnamed stadium worker in 2006 and by Juan Medina in 2014, and sound recordings I made of the stadium ambience in 2007. Read and Wyndham 2007, 'Between the Silence and the Scream', in Bandt and Duffy 2007.

6 Their published words give them the agency of which they were so brutally deprived and provide the backbone of this chapter.

7 The sound can be auditioned on the CD enclosed in Bandt, Duffy and McKinnon 2007, pp. 35–48. Read and Wyndham 2007, 'Between, in Bandt and Duffy 2007.'

8 Many interviewees remember the blood, urine and faeces in the corridor, making it very slippery.

9 In fact, soldiers were drilling holes in the bathroom tiles to stabilize the *parrillas*, or electrified metal grills used to torture detainees. The holes were unnoticed and their purpose unrecognized until identified by a survivor visiting the stadium in 2012.

10 In fact, they were already scheduled for execution. Their bodies were mostly thrown into the Mapocho River that flows through the city, or taken for disposal to the infamous Patio 29 of the Santiago General Cemetery. Nobody knows how many detainees were 'disappeared' from the Stadium, though some eyewitnesses estimate at least 100.

11 Godoy, Minister for Employment and Pensions; Godoy, Minister for Housing; Mario Céspedes Gutiérrez, leading left-wing academic and Congress member.

12 At this point, Jara started scribbling out his untitled but probably most famous composition, generally known as 'Estamos Cinco Mil' (We Are Five Thousand). The paper was smuggled out of the stadium inside a sock.

13 The well-known Communist party member and Director of Santiago's prisons; Nena Gonzales, interview.

14 The best contextual translation might be 'Mummy, mummy.'

15 That is, to Patio 29, Santiago General Cemetery. To this day, nobody is certain how many bodies were thrown into the excavated graves, nor of their identities. The Patio 29 caretaker, Nena González, estimates the number as high as 1000. Nena Gonzales, interview.

16 Most accounts refer to Jara's execution occurring on the final day, his spine and many bones broken, following which several soldiers were invited to put as many bullets into him as they liked. Possibly 40 shots were fired into his body.

17 Koestler 1937/1966.

References

Aránguiz, T. V., 2006. *Mujeres en Rojo y Negro*, Santiago: Ediciones Escaparate, pp. 302–14.

Bandt, R., Duffy, M. and McKinnon, D., 2007. *Hearing places*, Cambridge: Cambridge Scholars.

Cuevas, J. P., 2009. 'Los estremecedores testimonios de cómo y quiénes asesinaron a Victor Jara', ('The alarming testimonies of who killed Victor Jara and how'), 26 May. Available at http://ciperchile.cl/2009/05/26/los-estremecedores-testimonios-de-como-y-quienes-asesinaron-a-victor-jara/.

Fuentes, M., Sepúlveda, J. and San Francisco, A., 2010. 'Towards an archaeology of the Víctor Jara Stadium, massive detention and torture centre of the Chilean dictatorship'. Available at http://www.researchgate.net/profile/Miguel_Fuentes6/publication/259564561_Hacia_una_Arqueología_del_Estadio_Vctor_Jara_Campo_de_detención_y_tortura_masiva_de_la_dictadura_en_Chile_(1973–1974)/links/00b4952c8adaa2afcc000000.pdf.

González, Nena, interviews with the author, 2007, 2014.

Ida, Ruz de las Paños, 2007. *La Funa de Víctor Jara*, DVD.

Jara, J., 2001. *Un cantado truncado* (An unfinished song), Madrid: Punto de Lectura.

Koestler, A., 1937/1966. *Dialogue with death*, London: Pan Macmillan.

Luz, C., 1993. *La Flaca Alejandra* (documentary).

Medina, Juan, interview with the author, 2014.

Navia, B., n.d. 'El ultimo aliento de Victor Jara' ('Victor Jara's last breath'). Available at http://www.agenciaelvigia.com.ar/ultimo_aliento.htm.

Peña, J. C., 2008. 'La sangre de un poeta', blog, 15 October. Available at http://periodismoyexcelencia.blogspot.com.au/2008/10/juan-cristobal-pea-la-sangre-de-un-poeta.html.

Quiroga Cavajal, L., n.d. 'Memoria viva'. Available at https://www.google.com/search?q=litre+quiroga+chile&ie=utf-8&oe=utf-8, last modified 2010.

Read, P. and Wyndham, M., 2007. 'Between the silence and the scream: Recordings made at sites in the last days of Victor Jara' (with M. Wyndham), in R. Bandt, M. Duffy and D. McKinnon eds., *Hearing places*, Cambridge: Cambridge Scholars, pp. 35–48.

Read, P. and Wyndham, M., n.d. 'Narrow but endlessly deep: The struggle for memorialisation in post dictatorship Chile', Monograph MS. Under Review.

'Victor Jara and the story of his last song'. Available at http://www.mahmag.org/english/worldpoetry.php?itemid=380.

Villegas, S., 2013. *El Estadio: Once de septiembre en el país del Eden (The Stadium: The eleventh of September in the country of Eden)*, Santiago: Ediciones Septiembre, LOM.

Zuñiga, A. S., 'Testimonio de Antonio Zuñiga Senra'. Available at http://memoriaviva.cl/testimonios/testimonio_de_antonio_zuniga_senda.htm.

Part III
Sensory Memories

11 "Big Smoke Stacks"

Competing Memories of the Sounds and Smells of Industrial Heritage[1]

Lisa Murray

Redfern, Alexandria and Waterloo are inner city working class industrial suburbs located just to the south of Sydney's city centre. For over 100 years, from the 1850s to the 1950s, the area was the industrial powerhouse of metropolitan Sydney. Thousands of workers converged on the suburbs and the factories hummed and hissed to the sounds of industrial processes. Since the 1960s, industry has shifted further out from the city, and the inner suburbs shifted from heavy manufacturing to warehousing. Today the southern industrial area is undergoing radical urban renewal through residential re-zoning and state government intervention, with large apartment blocks being built on former industrial sites. The industrial landscape is now largely obliterated. The remaining working class residents are experiencing social dislocation in their once close-knit communities and are dwindling in numbers, replaced by a high-income population that wishes to live close to the city.[2]

This chapter charts the sensory experience of living in these once industrial working class suburbs and considers how the process of gentrification is shifting the social values and meanings of the 'big smoke stacks' that once dominated the landscape. Sensory experiences of place can help to historicize a community's environment and their identity with place. The shifting memories and meanings of sensory experiences are an underutilized source that can be used to chart the creeping impact of gentrification in late twentieth-century Sydney. The chimney stacks of Waterloo, which were an accepted part of industrial life for over 100 years, became in the 1980s an unwanted symbol of the pollution suffered by the community. However, as the factories were demolished and the stacks disappeared, those stacks that remained dormant and sentinel in the landscape began to take on new meanings, to be celebrated by newer residents as industrial chic heritage, a 'love song to the gritty city.'[3] Historians of public history and memory readily recognize that industrial heritage sites are often sites of contested memories. The role of memory and history in the creation of public memorials to a disappearing past is often hotly debated at a policy and political level, as well as by the community that suffers both economic loss and a loss of identity through deindustrialization. The unmaking and remaking of places through community memories is an integral, yet intangible, aspect of gentrification in the post-industrial city. It is an aspect of Sydney's history which needs to be explored further.[4]

The Whirr and Clatter of Nineteenth-Century Industry

Two geographical qualities primed Redfern, Alexandria and Waterloo to be a hub of noxious trades and manufacturing: the area's proximity to town, where the markets, warehouses and ports were all located, and its vast amounts of water. Collectively known as the South Sydney area, the municipality of Redfern adjoined the southern city boundary, with the municipality of Waterloo lying to the southeast of Redfern, and Alexandria municipality to the southwest. The main road from the city to Botany Bay passed through the area. This area, less than a mile south of the city, was once a series of vast undulating sand dunes covered by heath, low scrub, creeks and fresh-water wetlands.

The area's industrial development was given a push when the government banished all noxious trades from the city boundaries. The *Sydney Slaughter Houses Act 1849* gave the proprietors of tanneries, fellmongers, wool scour-ers, abattoirs, boiling-down works and soap factories ten years to move out of the city. Colonial industries and an expanding residential population from the 1850s both placed great stress on the water supply and natural drainage of Redfern, Alexandria and Waterloo. The industries brought with them other forms of pollution too: noise, smoke and disagreeable smells. As the population grew, so did complaints about the pollution. The healthful sea breezes and clear waters, once a much celebrated part of the district, became a talking point for Sydneysiders for all the wrong reasons.[5]

Coal burning furnaces spewed smoke and soot into the atmosphere from tall brick chimneys. Gases and smells drifted from different manufactories according to the processes being undertaken. Sometimes the smells were sickly sweet, like the hops, yeast, sugar or fruit smells which emanated from the breweries, bakeries and jam manufactories. Other times they were gas-eous or unpleasant, like the tallow and paraffin smells from fat extractors and candle manufactories.

Rarely was there a let-up. Goodlet & Smith's kilns at their Waterloo brickworks and the retorts at the kerosene refinery in Waterloo both oper-ated continuously. Other processes might have been more intermittent, but all machinery required furnaces and steam engines. Those living close to manufactories sometimes had to shut up the house to stop the smoke get-ting in. Airborne soot was a common nuisance for all, soiling linen and dirtying households.[6]

Unpleasant smells from the tanneries, candle and tallow works were part and parcel of the southern suburbs. Richard Seymour, Inspector of Nui-sances at the Sydney City Council, claimed no amount of whitewashing and cleansing could abate the smell of boiling-down establishments.[7] It was all-pervasive. Industrialists and inspectors agreed; the 'soup' in which the bones and meat were boiled was the nuisance. Rancid fat and meat added to the stench; so too the piggeries and duck farms that consumed the leftovers.[8]

The pungency and prevalence of the smells from these trades were exacerbated by inappropriate drainage. The soups and soakage were run-off into open surface drains or into the local creeks. In the worst cases, they were poured onto the ground to simply drain through soakage and evaporation.[9]

An 'out of sight, out of mind policy' prevailed. Throughout most of the nineteenth century, distance was seen as the only cure for nuisances.[10] That was why slaughter yards and noxious trades were pushed outside the city boundaries in the first place. Few leaders, at either a municipal or colonial government level, felt much sympathy or concern for the residents on the sand hills of South Sydney. Nuisances were seen as an inevitable part of the trades and, as the trades were integral to the 'progress' and economic prosperity of the colony, there was nothing to be done except remove them away from densely populated districts.

Unfortunately, the suburbs of bountiful water were to the south of the city, and the offensive smells emanating from the oil refinery, tanneries and boiling-down works wafted towards town on the prevailing southerly winds. The colonial legislators should have been cognizant of this problem. The 'southerly burster' was a regular weather phenomenon, part of a warm Sydney summer's day. The 'brickfielder,' which swept dirt and dust from the brickfields immediately to the south of the town, had been annoying colonists since the 1800s. And the shifting sands of Strawberry Hills and Surry Hills had threatened to engulf the growing town in the 1850s. But all this was forgotten, or at least ignored, when our statesmen banned noxious trades from the inner city in 1849.

The boiling-down establishments made themselves felt in the warmer, humid months. Smells from tallow works hung around on still evenings and were particularly noticeable as the first glimmers of dawn appeared. Nuisance Inspector Seymour reported, 'If you pass along the Cook's River Road in the morning about 4 o'clock, and between that time and 5, you can smell this boiling-down all along the road.'[11] The stench made people's stomachs turn. Occasionally people on the passing omnibus vomited out onto the street.[12]

Changes in wind direction brought different smells, but rarely a let-up in unpleasant odours. The kerosene works in Waterloo could be distinctly smelt in Randwick on the tail of a southerly wind. The northeast sea breeze sent smells across Redfern and Waterloo towards Alexandria and Macdonaldtown.[13] Redfern nurseryman John Baptist reported that when he wanted to enjoy the northeast breeze from the coast, his gardens were infiltrated by the offensive smoke from the chimney at Alderson's tannery. When a cooler southerly change came along, Baptist smelt the effluvium from the drain below Magill's and his property. Here 'accumulated fluid filth' from the tannery became 'a pond covered with a frothy and thick slimy scum.'[14]

Distance may not have obviated the smells, but it did cast away the incessant industrial noise. It was fortunate that the NSW Shale and Oil Refinery was isolated down in the swamps of Alexandria, as the 'wild whirr' of the machine bands, and the 'deafening clatter' of the agitators, kept up 'an unceasing din.'[15]

Noise and clatter were the celebrated sounds of industrial progress in nineteenth-century Sydney. Alderson's six-acre tannery site in the expanding suburb of Redfern bustled with activity. Newspapers admired the manufactory: 'with its tall chimney and its clanking engine [the tannery presented] an appearance of busy life that one would hardly expect to meet with amongst the sand hills.'[16] The steam engines reverberated, causing an audible and physical nuisance. Alderson's next-door neighbour, Michael Magill, complained of vibrations and shaking so hard that one of the walls on his property fell down.[17]

The Rhythm of Life: Living beside the Factories

Land opened up in the twentieth century for more intensive industrial development. The aldermen of Alexandria and Waterloo welcomed industry with open arms. The noxious trades, railway workshops and goods rail yard were joined by heavy manufacturing, food processing, furniture manufacturers and chemical factories.[18] Oral history interviews conducted by Sue Rosen in 1994–95 with long-term South Sydney residents confirm that the industrial smells and shift work of the factories were part and parcel of the urban neighbourhood throughout the twentieth century. Workers arriving by train and tram swept through the suburbs, walking to major employers such as Metters' and Austral Bronze. Metters' was a household name, as the largest stove-maker in Australia. At their peak, they employed 2500 staff at their 28-acre Alexandria factory. Austral Bronze was just down the road from Metters', in Mitchell Road in Alexandria. The Austral Bronze Company was formed in 1915 to manufacture brass ammunition feeds for machine-gun belts. After World War I, the company added rolled brass and copper to their manufacturing line up. At its peak, the company employed 1600 workers.

The factory whistles, audible across the suburbs, signalled the changes in shifts. 'People conducted their lives by the railway whistle,' said Harry Brennan, who grew up in Alexandria in the 1950s and 60s. 'The seven-thirty whistle in the morning, the four o'clock whistle in the afternoon. And then there was an eleven o'clock whistle of a night at the glass factory at Waterloo.' The whistles became the suburban timepiece. They were part of the rhythm of life.

The smells of the tanners and the boiling-down works remained a signature of the district in the twentieth century and impregnated the memories of locals. Stephen Fennell grew up in Alexandria in the 1950s and distinctly remembered the boiling-down factories. 'Look, those smells—[what] I do remember is the bloody meat place out at Mascot [and] the soap manufacturing place. If that wind, you know, turned and come off the south-east, yes, you could smell that. But those sorts of things have been overcome. Can't think of the names, Bird Brothers I think they were. But they, you know, used to clip all the bones from the butchers shop, and make the soap and that and you could smell that.'[19]

Smoke and soot tainted the air. Harry Brennan has never forgotten the smell of the railway workshops at Eveleigh. The government railway workshops on the western side of Redfern were the heart of railway transport in New South Wales and became one of the largest industrial sites in Australia. Redfern's economic and community life became intertwined with the Eveleigh government railway workshops. As much as possible was manufactured at the workshops, but many other items were obtained from local businesses and private contractors, following the departmental policy to buy Australian-made. The workshops employed 3260 people in 1912, from boilermakers and blacksmiths, coachmakers and painters, to cleaners and clerks. 'There was a certain smell about Alexandria . . . an industrial smell. . . . Some days it was more so, you know. Some days you wouldn't hang the washing on the line either, because [of] ash [and] soot.' The hissing and puffing of the steam engines at the goods yards and workshops were familiar sounds. The Alexandria Goods Yard managed country and interstate carriages, so there was the constant movement of trains being shunted, loading up coal, getting up enough steam to drive the engine or letting off steam.[20]

Smoke and soot did not just come from the railway workshops. The skyline was a forest of chimney stacks. Boilers, incinerators, fires and smokestacks belched soot and smells across the district, polluting the air and soiling washing. Complaints to authorities and politicians were effective only occasionally, because the side effects of industry were an unavoidable fact of life for anyone living in the area.[21]

The food manufacturers added sweeter, cloying smells. The smell of hops fermenting in the Resch's Brewery was not particularly popular, but the chocolate, jam and sauce factories filled the air with sickly sweet perfumes. Francis Chocolates in Stirling Street, Redfern was just behind Bev Karonidis' childhood home. 'You used to always know when they were making caramels and things like that,' said Bev, 'because you would always smell all that sort of thing.' The IXL jam factory smelt of pineapples, melons and tomatoes: 'When they were making tomato sauce it was beautiful.'[22]

Anne Ramsay, who was born in 1914 and lived in Redfern all her life, remembers the skyline was a forest of chimney stacks. By the 1990s, the factories were disappearing. 'Well there were more factories going before than what there is now. Because you see the brewery had the big smokestacks. Reckitts Blue had the big smokestacks, the bakery had the big smokestacks, see there is none of those now, because they've all been pulled down.' It was a sign of the de-industrialization of the suburb.[23]

As the area became more gentrified, people began to complain about the pollution. Bev Karonidis was born in 1936 and lived in Redfern all her life. Just down the road from her home was an iron foundry. It had been there all her life, but when she was interviewed in 1994, she had just heard that the foundry was closing the next week. As she put it, 'Because see, the newer people coming in complain about it. Which annoys me, because they . . . it was there before them.' Bev acknowledged that the foundry belched smoke at different times. 'Years ago it used to be a bit dirty one day a week, they used to have black stuff come out, smoke and that, and you wouldn't put your washing out. But they don't now, and we don't object.' It was accepted as part of the working life of the area. 'I mean the place was more alive. We had factories, not a lot, but there were a few of them around. You'd have working people and that. . . . But also you knew the people that worked. I mean the chaps at the foundry, they would know you.' The newer people moving into the area—the musicians, artists and designers—complained about things, according to Bev, created parking problems and closed things down.[24]

Gentrification, and newer residents' complaints about pollution, certainly played a part in the deindustrialization of the area. But also people's understanding of and perceptions about pollution were also changing. Reflecting in 1994 upon her childhood, Anne Ramsay remembered the 'dirt and the squalor' growing up in Redfern, but she never really thought it was polluted. 'The front verandahs [would] be filthy, but you wouldn't think of it as being putrid [or polluted]. We had never heard of pollution then. It's only in the later years that we had heard it called pollution.'[25] Anne Ramsay's comment highlights the importance of 'historicizing' the senses. The reception of sensory experiences in the past is not the same as today. In this case, the shift in perception from dirt to pollution reflects a historical moment in time, in the 1970s and 1980s, when urban gentrification was infiltrating the inner city suburbs.

Gentrification Begins

The area encompassing Redfern, Alexandria and Waterloo has been described by some real estate agents as the last frontier of gentrification in the inner city. The area's low socio-economic status, the high proportion of public housing, negative publicity about 'The Block' (the colloquial name for a block of Aboriginal housing in Redfern bounded by Louis, Vine, Eveleigh and Caroline streets and owned by the Aboriginal Housing Company), the high incidence of drug abuse and crime, and constant negative media attention combined in the late-twentieth century to create a public perception of the area as unsafe, a no-go zone. Nevertheless, the area's demographics and outlook have been slowly shifting since the 1970s.

Redfern, Alexandria and Waterloo were at the tail end of successive waves of inner city gentrification. Until the 1980s, the area was steadfastly working class, dominated by blue-collar workers with low socio-economic status and a high proportion of public housing. As pressure on housing prices grew in suburbs nearby, Redfern, Alexandria and Waterloo became increasingly desirable to middle-income professionals, and low-cost private rental accommodation began to be squeezed out. Increases in the number of white-collar workers living and working in South Sydney are noticeable from the 1970s, and there was an 84 per cent increase in the number of managers, administrators and professionals living in the South Sydney area between 1986 and 1996. The change in housing costs is even more astounding. Median house prices in the South Sydney area between 1982 and 1995 increased roughly 10 per cent each year, and median unit prices increased 12 per cent a year over the same period.[26]

The process had political consequences. The rise in civic and resident action groups agitating for improvements from the 1970s is closely associated with gentrification. There was growing concern in the district about public health and safety from hazardous industries in the 1970s. Increased traffic congestion, the transport of hazardous waste through the municipality and air pollution were topics hotly debated by residents and community groups such as the Inner Sydney Regional Council for Social Development. The Save Our Suburbs campaign mobilized South Sydney residents to protest against traffic corridors, increasing congestion and pollution. Arty types and creatives moved in and started complaining about the pollution caused by the remaining factories. What was once just 'dirt' now became 'pollution.'[27]

The public housing campaign in Waterloo was another sign of the gradual gentrification of the area. While primarily airing the views of long-term private and public housing tenants, the campaign benefited from the drive and knowledge of women such as Margaret Barry, who had bought a house in the area and helped set up the resident action group. The No Third Runway campaign was yet another example of the changing attitudes in the district.

The Zetland Monster: Ban the Burn Campaign

The shift in public attitudes towards industry and pollution, and its intersection with the gentrification process, is best represented in the fight to close the incinerator in Zetland, a small suburb next to Waterloo. Bad odours, soot and fall-out, smoke and air pollution had been part and parcel of living in Alexandria and Waterloo for well over 100 years. But grass-roots politics, which fostered resident action groups, growing environmental awareness and increased community empowerment to express a desire for improved public amenity meant that by the 1970s industry was no longer seen as the saviour of the district. Add a little bit of gentrification into the mix, and the 20-year campaign to close the incinerator became a cause celebre.

The modern incinerator, called the Waverley Woollahra Processing Plan after the councils which invested in it, was officially opened along Bourke

Street, Zetland by the NSW Governor, Sir Roden Cutler, on 27 June 1973. The garbage incinerator operated 24 hours a day, seven days a week, with a staff of 32 people. It processed household and trade refuse from all the surrounding council areas: Waverley, Woollahra, Sydney City, South Sydney, Randwick and Botany—up to 180,000 tonnes of waste each year.[28]

Residents of Zetland, Waterloo and Beaconsfield objected to the incinerator right from the start. Visible fall-out and smells from the plant fuelled community concerns about toxic emissions. Within a year of the incinerator commencing operations, South Sydney Council was receiving complaints from residents. A special public meeting was convened in Redfern Town Hall in August 1974 to air residents' concerns and protest to the state government about the detrimental effects of the incinerator.[29]

The Mayor of South Sydney, Alderman Vic Smith, was a vocal critic of the incinerator. 'For years the residents of South Sydney have had to suffer smoke pollution, waste fall-out, noise and odour emissions,' he said in 1992. 'Of recent times they have had to endure the increasing emission of toxic substances, notably dioxins and furans.'[30]

Ann Brown was one of the new residents who joined the campaign against the incinerator. She was a relative newcomer to Zetland, part of the new wave of gentrification that took hold of the suburb in the 1980s. Ann was a trained nurse and she bought her house in Portman Street because it was close to her work and close to the city. Her home was still affordable and the area was not too crowded. The downside—as she discovered—was that she had an incinerator at her back door.

An application by Waverley Council to expand the processing plant and establish a depot beside the incinerator galvanized the local residents. 'It began to dawn on us,' Ann said, 'that we were going to be the dumping ground for all the eastern suburbs' trash, with no say . . . we felt that if we weren't careful that this whole area with all these redundant industrial sites would be taken over as cheap land for anybody that wanted to do something on the site.'[31]

The putrid smells worsened, and traffic movements increased through the 1980s and 90s. Investigations by the State Pollution Control Commission in 1991 put a spotlight on the plant's operations. The Commission issued a warning regarding unacceptable air pollution—it had recorded dioxin levels 27 to 153 times the overseas standards.[32] Public attitudes hardened against the incinerator as the environmental and public health risks were publicized. 'Once we began to find out the facts it took on a huge momentum because it became clear just what the health risk was, that there was no social justice in this at all, that it wasn't even owned by our council.'[33]

The year 1992 marked a turning point. In August 1992 the environmental group Greenpeace occupied the incinerator site for 36 hours, brandishing placards and gas masks. It brought much-needed publicity to what Ann described as a 'David and Goliath struggle.'[34]

The regulatory environment also changed in 1992, with the introduction of the *Protection of the Environment Administration Act*, creating the Environment Protection Authority. Woollahra and Waverley councils faced uncertainty over environmental regulation. With the facility running at capacity and potentially requiring a significant upgrade to meet new standards, the future of the incinerator was hanging in the balance. The owners decided to upgrade the facility.

The Zetland Community Action Group was formed by residents to coordinate community protest and resistance to the incinerator. The 'No Incinerator' campaign was supported by South Sydney City Council and was officially launched at the South Sydney Festival in October 1992. Badges, car stickers and posters proclaiming 'No Incinerator' were plastered across Zetland, Beaconsfield and Waterloo.[35]

The community kept up the pressure through the 1990s against what locals derogatorily called the 'Zetland Monster.' National groups such as Greenpeace and local groups such as the Inner City Regional Council for Social Development kept the issue in the public domain with fliers, lobbying and protests. The 'Ban the Burn' rally against the Waterloo incinerator was held in September 1994 and in November, Greenpeace and the Zetland Community Action Group established a 'dioxin hospital' on the incinerator's grounds.[36]

The incinerator finally closed in December 1996. A community celebration was held in the residential streets surrounding the incinerator to mark its demise. The site remained vacant as Waverley and Woollahra councils negotiated the sale of the site, a prime candidate for urban renewal. The closure of the incinerator breathed new life into the community. Residents had won their David and Goliath battle, sending a message that the area's many vacant industrial sites could not become a dumping ground for Sydney's toxic waste. Local resident Ann Brown described it as 'a psychological shift, that people did feel the area had a future.'[37] However, there was a human cost to the campaign. Ann Brown saw 'idealistic people who believed in the truth become burnt out and angry.' While the incinerator's closure was beneficial to the area's health and lowered toxic pollution levels, it also brought the juggernaut of urban renewal that bit closer.

Industrial Chic

The now idle chimney stack stood sentinel over the community's victory, but its public meanings began to shift as it became emblematic of the deindustrialization occurring across the district. Deindustrialization paved the way for a third wave of gentrification, which began from the 1990s, driven by state and local government as an urban development strategy. It involved the adaptive reuse of old buildings for commercial and residential development, as well as the building of new ones. The closed Eveleigh Rail Yards were partially redeveloped by the government in the 1990s into Australian

Technology Park, a high-tech commercial venture with links to the universities, and the contemporary performance space Carriageworks opened in 2007. This signalled the start of urban renewal in the district.

The former industrial spaces of Redfern, Alexandria and Waterloo were perfect brown-field sites (sites previously used for industrial purposes) for private developers and public–private partnerships. Among the earliest developments was the conversion of McMurtrie's shoe factory into the Watertower apartments. Factories were being demolished, replaced by new-build master-planned apartments, including the building of the Moore Park Gardens apartments on the former Resch's Brewery site and Meriton's Crown Square on the former ACI Glassworks site.

Landcom's public–private partnership of the former BMC–Leyland factory site brought another 2500 residential apartments into Zetland. This brown-field site was transformed into modern apartments and landscaped open spaces. The name, Victoria Park, is a reference to the site's history as a proprietary racecourse. The development's innovative design, architecture and environmental features have garnered 16 industry awards. The success of the public–private development signalled the possibilities for Green Square.[38]

The retention of decommissioned machinery and tall chimney stacks within public spaces has been the chief form of memorializing the disappearing industrial history of Sydney. Sydney Park retains some of the kilns of Bedford Bricks; a stack from the ACI glassworks site in Redfern is enveloped and almost overshadowed by modern apartments; a single crane perches on the foreshore at Blackwattle Bay beside the new walkway. This phenomenon is not, of course, unique to Sydney. The contested terrain of industrial heritage within a post-industrial landscape is being encountered across Australia, Britain, Europe, Canada and the United States of America.[39] The preservation of such items is driven by aestheticism and a quest for economic cultural tourism or urban place-making more than any historical understanding of the economic, political or social experiences of industrial workers. The memories and experiences of the local community and the industrial workers are entirely absent from such forms of landmark memorialization of industrial life. Even in the process of retaining items of industrial heritage, the history of the local place and its community is being obliterated.

As memories of the incinerator's pollution and the long-running campaign to shut it down faded from wider public consciousness, the Zetland Monster's chimney stack began in the early 2000s to be admired for its industrial beauty. Without the smoke, the chimney was a benign structure that could become a landmark in urban place-making. A new form of historical consciousness was being attached to the big smoke stack's industrial chic heritage. The pollution was being forgotten, and the majesty of large-scale industrial processes was being remembered. Architects and heritage planners began airing the viewpoint that the chimney should be retained and the incinerator adaptively reused. The late Col James, an architect and community campaigner, wanted the building salvaged and reused to provide cheap accommodation for artists and students. 'Buildings should be recycled not destroyed,' he argued in the *South Sydney Herald* in February 2008. 'It's a great tragedy. It's one of the best things that Green Square has going for it.'

Conclusion

The Zetland incinerator became a site of contested memories. The incinerator and its chimney were finally demolished in 2008. Local campaigners and those affected by the emissions were glad to see it go. Local resident and Waterloo businessman Col Charlton told Sarah Malik of the *South Sydney Herald* in February 2008, 'I'm glad to see it go. It's an ugly thing anyhow. It should have never been there in the first place—certainly it should not have been in the middle of the city.' Trevor Davies, editor of the same local rag, bid the structure good riddance in an editorial in April 2008:

> Standing at a bus stop on Botany Road a few weeks ago I was pleased to see that work on demolition of the Waterloo incinerator had started. I, for one, am pleased. I was working as a street sweeper in the Redfern Waterloo area in the early 90s and my health was affected—my asthma became a problem. I always thought what cheek two Eastern Suburbs Councils—Woollahra and Waverley Council—had in burning their rubbish in Redfern Waterloo and not in their own back yards. I mean, why not take it to the North Shore, maybe somewhere like Lindfield, and build an incinerator? No, they chose [working class] South Sydney. I'm sure I speak for many people when I say that it should have been torn down years ago.

As the urban renewal plans for the Green Square town centre at Zetland move from master planning to reality, more people are expressing regret that the incinerator was not retained in some way. This shifting public memory of the incinerator post-closure is part of the process of gentrification and urban renewal that is occurring in the district. Sites of industrial work and heritage will continue to harbour contested memories for dislocated residents. For those who never experienced the incinerator with its putrid smells, who never had to dodge the garbage trucks, who never had their washing covered in soot, or felt the vibrations of the incinerator, the chimney stack was simply a majestic reminder of the area's industrial past that was being obliterated.

But the chimney stack could never remain. For the local community it was the 'Zetland Monster.' It had been a threat to their health for over 20 years. They campaigned relentlessly for its closure. In many respects, if it was not for them, the whole urban renewal and creation of the Green

Square town centre could not have occurred. As I deal with the history of this place, and advise on heritage interpretation as part of the urban place-making being undertaken by the city council, I am becoming increasingly ambivalent about how the 'Zetland Monster' should be interpreted. Col James was right. The chimney was one of the few things Green Square had going for it, particularly in terms of urban place-making. But perhaps, just sometimes, a piece of industrial heritage should be obliterated to respect the social values and history of the community. It is the story of the community and their campaign for the closure of the incinerator that needs to be told, rather than a celebration of 'big smoke stacks.'

The history of the industrial city and our understanding of community lives are enhanced when we focus on the history of the senses. The intimacy of sensory experience forces the historian to consider the everyday embodiment of lives, past and present, the networks of community connections and shared experiences. A sense of place and community identity are not defined by maps and council boundaries but by social groups, neighbourhoods and shared memories. The emotional connection and memories conjured by the senses are an intangible yet integral part of every community. By historicizing memory and the senses, we can begin to chart the social impact of the changing urban environment and demographics, such as deindustrialization and gentrification on inner city working class communities.

Notes

1 This chapter draws upon research undertaken by the author while in the position of City Historian at the Council of the City of Sydney.
2 Hayes 1985; Thompson 2011.
3 Farrelly 2006.
4 An Australian perspective is provided by Gregory 2008. European and North American studies include Frisch 1998; Paz and Gonzalez 2012; Judson 2014; Twigge-Molecey 2014.
5 New South Wales Legislative Assembly 1875–76b; Fitzgerald 1987, pp. 62–63.
6 New South Wales Legislative Assembly 1862: Evidence: Mr R. A. Hunt Q.619–25; Sydney Morning Herald 1863; Australian Town and Country Journal, 1875; Australian Town and Country Journal 1877.
7 'Report of the Royal Commission of Inquiry into Noxious and Offensive Trades', *New South Wales Legislative Assembly Votes and Proceedings*, 1, 2nd session (1883): Evidence: Mr R. Seymour Q.22.
8 New South Wales Legislative Assembly 1883: Mr R. Seymour Q.137–43, Mr. J Walsh Q.720–24.
9 This is clearly articulated in all the evidence in the Tanners and Curriers Bill Report (New South Wales Legislative Assembly 1862) and also the Noxious and Offensive Trades Royal Commission (New South Wales Legislative Assembly 1883).
10 'The only alternative is to trust to distance to keep nuisances at bay', New South Wales Legislative Assembly 1883, p. 16.
11 New South Wales Legislative Assembly 1883: Mr R. Seymour Q.22.
12 Ibid., Mr. A. R. Fremlin Q.2812. Similar reactions were described by the medical practitioner Isaac Aaron in relation to emanations from tanners: New South Wales Legislative Assembly 1862.

13 New South Wales Legislative Assembly 1883: Mr W. Bradley Q.2652, 2685–90; Sydney Morning Herald 1877.
14 Alderson Case and sewerage problems in Redfern, 1876, Baptist Family Business Records, 1832–1921, including those of The Gardens and Surry Hills, together with personal papers, ML MSS 162, box 24 [5], State Library of New South Wales; New South Wales Legislative Assembly 1875–76a: Evidence: Mr J. Baptist Q.72.
15 Australian Town and Country Journal 1875.
16 Illustrated Sydney News and New South Wales Agriculturalist and Grazier 1880; Sydney Morning Herald 1865.
17 New South Wales Legislative Assembly 1875–76a: Evidence: M. Magill, Q.109–10.
18 Waterloo Municipal Council 1920; Alexandria Municipal Council 1943.
19 City of Sydney oral histories: Stephen Fennell.
20 City of Sydney oral histories: Harry Brennan, Keith Mulhearn, Stephen Fennell.
21 City of Sydney oral histories: Anne Ramsay, Cathi Joseph, Mick Green.
22 City of Sydney oral histories: Anne Ramsay, Bev Karonidis, Claire Mulhearn, Keith Mulhearn, Jill Edwards.
23 City of Sydney oral histories: Anne Ramsay.
24 City of Sydney oral histories: Bev Karonidis.
25 City of Sydney oral histories: Anne Ramsay.
26 Hayes 1985, pp. 153–54; Murphy and Watson 1990; South Sydney Council 1999, pp. 46–47; Spearritt 2000, pp. 181, 203–04; Thompson 2011, pp. 18–19.
27 City of Sydney oral histories: Anne Ramsay, Bev Karonidis.
28 Humphreys, n.d. p. 8; Maunsell Pty Ltd 1992.
29 South Sydney Municipal Council 1974a and 1974b.
30 Sydney Morning Herald 1992.
31 Ann Brown in Beasley 2001, p. 29.
32 Humphreys n.d., p. 9.
33 Ann Brown in in Beasley 2001, p. 29.
34 Ibid.
35 Sydney Morning Herald 1992.
36 Green Left Weekly n.d.
37 Ann Brown in in Beasley 2001, p. 29.
38 Residential Developer Magazine 2009.
39 See for example Hayden 1997; Frisch 1998; Mansfield 2005; Paz and Gonzalez 2012.

References

Alderson Case and Sewerage Problems in Redfern, 1876. Baptist Family Business Records, 1832–1921, Including Those of the Gardens and Surry Hills, Together with Personal Papers, ML MSS 162, box 24 [5], State Library of New South Wales.

Alexandria Municipal Council, 1943. *Alexandria, 'The Birmingham of Australia': 75 years of progress, 1868–1943*, Sydney: Alexandria Municipal Council.

Australian Town and Country Journal, 1875. 'The NSW Shale and Oil Company's refinery', *Australian Town and Country Journal*, 10 April, p. 573. Available from http://nla.gov.au/nla.news-article70490126.

Australian Town and Country Journal, 1877. 'Messrs. Goodlet and Smith's annular kilns Waterloo', *Australian Town and Country Journal*, 3 February, p. 188. Available from http://nla.gov.au/nla.news-article70598643.

Beasley, M., 2001. *Everyone knew everyone: Histories and memories of Green Square*, Sydney: South Sydney Development Corporation.

City of Sydney Oral Histories: Harry Brennan, Jill Edwards, Stephen Fennell, Mick Green, Cathi Joseph, Bev Karonidis, Claire Mulhearn, Keith Mulhearn, Anne Ramsay. Available from http://www.sydneyoralhistories.com.au/.

Farrelly, E., 2006. 'Love song to the gritty city', *Sydney Morning Herald*, 31 March, p. 15.

Fitzgerald, S., 1987. *Rising damp: Sydney 1870–90*, Melbourne: Oxford University Press.

Frisch, M., 1998. 'De-, re-, and post-industrialization: Industrial heritage as contested memorial terrain', *Journal of Folklore Research*, 35(3), pp. 241–49.

Green Left Weekly, n.d. '30-year battle to ban the burn', *Green Left Weekly*, (243), Available from https://www.greenleft.org.au/node/11193.

Gregory, J., 2008. 'Obliterating history? The transformation of inner city industrial suburbs', *Australian Historical Studies*, 39, pp. 91–106.

Hayden, D., 1997. *The power of place: Urban landscapes as public history*, Cambridge, MA: The MIT Press.

Hayes, H., 1985. *South Sydney community profile*, Sydney: South Sydney Committee of the Australian Assistance Plan.

Humphreys, A., n.d. 'Waterloo Incinerator—Historical Analysis & Assessment of Significance', Architectural History Services.

Illustrated Sydney News and New South Wales Agriculturalist and Grazier, 1880. 'Messrs. Alderson and Son's leather exhibits at the Garden Palace', *Illustrated Sydney News and New South Wales Agriculturalist and Grazier*, 21 February, pp. 2–3. Available from http://nla.gov.au/nla.news-article64973354.

Inner Sydney Voice: The Journal of the Inner Sydney Regional Council for Social Development, 2008. 30th anniversary edition, 112, Summer.

Judson, S., 2014. ' "I am a nasty branch kid": Women's memories of place in the era of Asheville's urban renewal', *The North Carolina Historical Review*, 91(3), July, pp. 323–50.

Mansfield, M., 2005. 'Reimagining the suburbs: An investigation of a place making strategy in a deindustrializing city', *Asia Pacific Journal of Arts and Cultural Management*, 3(1), pp. 179–87.

Maunsell Pty Ltd, 1992. *Planning for the future of the Waverley Woollahra process plant: Final report*, Sydney: Maunsell Pty Ltd September.

Murphy, P. and Watson, S., 1990. 'Restructuring of Sydney's central industrial area: Process and local impacts', *Australian Geographical Studies*, 28, pp. 187–203.

New South Wales Legislative Assembly, 1862. 'Report from the Select Committee on the Tanners and Curriers Bill', *New South Wales Legislative Assembly Votes & Proceedings*, 5.

New South Wales Legislative Assembly, 1875–76a. 'Sydney City and Suburban Sewage and Health Board, 8th Progress Report, Report of Committee No. 6—To Examine into the Condition of Sheas Creek', *NSW Legislative Assembly Votes & Proceedings*, 5.

New South Wales Legislative Assembly, 1875–76b. 'Sydney City and Suburban Sewage and Health Board, 9th Progress Report,' *New South Wales Legislative Assembly Votes and Proceedings*, 5.

New South Wales Legislative Assembly, 1883. 'Report of the Royal Commission of Inquiry into Noxious and Offensive Trades', *New South Wales Legislative Assembly Votes and Proceedings*, 1, 2nd session.

Paz, B.D.P. and Gonzalez, P.A., 2012. 'Industrial heritage and place identity in Spain: From monuments to landscapes', *The Geographical Review*, 102(4), October, pp. 446–64.

Residential Developer Magazine, 2009. 'Urban renewal by Landcom', *Residential Developer Magazine*, January, p. 63.

South Sydney Council, 1999. *South Sydney Social Issues Paper*, November.

South Sydney Municipal Council, 1974a. South Sydney Council Minutes (3 July 1974, p. 3940; 17 July 1974, p. 2956; 31 July 1974, p. 3979), City of Sydney Archives.

South Sydney Municipal Council, 1974b. Reports from Special Public Meetings of South Sydney Municipal Council, CRS 442, City of Sydney Archives.

Spearritt, P., 2000. *Sydney's century: A history*, Sydney: UNSW Press.

Sydney Morning Herald, 1863. 'Sydney municipal council', *Sydney Morning Herald*, 12 September, p. 7. Available from http://nla.gov.au/nla.news-article13084210.

Sydney Morning Herald, 1865. 'The manufacturing industry of New South Wales, IV—The leather trade', *Sydney Morning Herald*, 21 April, pp. 6–7. Available from http://nla.gov.au/nla.news-article13109089.

Sydney Morning Herald, 1877. 'Creating a public nuisance—The charge against Messrs. Alderson and Sons', *Sydney Morning Herald*, 16 January, p. 7. Available from http://nla.gov.au/nla.news-article13392583.

Sydney Morning Herald, 1992. 'S. Sydney to break out in rash of "no incinerator" posters', *Sydney Morning Herald*, 1 October, p. 83.

Thompson, C., 2011. 'Master-planned Estates in the Inner-city as an Example of New-build Gentrification: Social and Cultural Change and Governance', PhD thesis, Department of Environment and Geography, Macquarie University, 21 August.

Twigge-Molecey, A., 2014. 'Exploring resident experiences of indirect displacement in a neighbourhood undergoing gentrification: The case of aint-henri in Montreal', *Canadian Journal of Urban Research*, 23(1), Summer, pp. 1–22.

Waterloo Municipal Council, 1920. *Waterloo Jubilee 1860–1920: Free industrial exhibition, Town Hall Waterloo*, Sydney: William Brooks Ltd.

12 Intimate Strangers

Multisensorial Memories of Working in the Home

Paula Hamilton

This beautifully polished silver bell was brought out by my elderly interviewee, JM, in 2015.[1] We had been discussing her upbringing in a house with servants and she brought it out as a keepsake of this time over 80 years ago and in Perth, across the other side of the Australian continent from the interview. Written language is inadequate to the task of describing the surprisingly loud but sonorous tinkle of this bell which was used to call servants into the dining room, to clear plates from one course of a meal and to bring in the next one. Yet, to hear it, like listening to JM's account of growing up, was to bring the past into the present, to collapse the distance of 'long ago' and re-imagine it visually in both our minds as part of the exchange. JM was one of four daughters of a doctor who had young live-in maids recruited first by word of mouth, and later from St Joseph's convent, until the mid-1940s. Her memory is infused with nostalgia about the youthful time with her sisters.[2]

Though it might seem jarringly anachronistic in the twenty-first century, bells such as these were a widely understood symbol of being 'at the beck and call' of employers, and they were used in both British and Australian homes up to the 1950s. Bells signified that for servants the home was an important listening environment, what the scholar Kate Lacey calls an 'acoustic space which is a "resonant" sphere with no centre and no margins.'[3] Although 'active listening' is historically variable,[4] at this time, when servants moved through the house, they were constantly on the alert, in a state of anticipation to being called or summoned. There were also different registers of enforced listening. One maid employed at a doctor's house disliked answering the telephone because it would ring 'all hours of the day and night,' and she subsequently refused to install a telephone in her own home even in old age.[5] In large homes, a system of bell-pulls operated from each room to a board in one central place. Margaret Mackie, whose father was a professor of education at the University of Sydney and Principal of Sydney Teachers College grew up in Wahroonga, a well-to-do suburb of Sydney, where she was born in 1915. She remembers a household dominated by bells:

> In the kitchen was a bell indicator, showing which electric bell had been rung. There were bells in the dining and drawing rooms, and in the main

Figure 12.1 The beautifully polished bell used by JM's family to call servants.
Photograph taken by Paula Hamilton, 15 January 2015.

bedroom. At meals, the bell hanging over the table in the centre of the room was rung when the next course was required. At breakfast, the housekeeper and, later, the maids, would first serve porridge and then, when the bell was rung, bring bacon and eggs or boiled eggs. The bell would be rung to summon the person in the kitchen if something was needed which had been omitted when the table was set.[6]

It was not merely the constant sound of these bells which underlined unequal power relations between members of the household that is notable here (bells with this function are associated with acoustic urgency), but Mackie's description also reveals the highly regulated nature of everyday life in some mid-1920s homes.

This chapter explores some aspects of the sensory landscape in the family home during the first half of the twentieth century by women who experienced it, who remember sound, sight, smell and touch but who were not *of* the family. Residential domestic servants may often have characterized their treatment as being 'like one of the family,' but this speaks of remembering disparate power relationships with affection rather than a literal sense of belonging; for no family member was summoned by a bell nor ate in the kitchen alone. The memories of servants, 'maids' or 'domestics,' as they were often called by that time, provide us with particular insights into the lives behind closed doors: not just the routine and the drudgery which is well documented, but an historicized understanding of privacy now understood as a space where one is not observed by other people and social relations in the home. For we see from the sensual landscape that home was not a 'safe' retreat, the physical space and emotional refuge that it was often characterized. The presence of employed workers in that space signals an infinitely more complex negotiation of private and public, exterior and interior, in different parts of the house across time. Depending on the social relations, servants were both highly visible and completely invisible: they could be both allies or spies, helpers or interlopers, in the joint enterprise of living and working in the privatized household. (Although servants were more often than not employed to 'do the dirty work,' as it was called, there were some instances where the work was 'shared.') Several scholars have argued that the central pillar of 'private life' at this time was understood as the heterosexual family and the state had made 'efforts to stabilise boundaries between private and public by mooring it to the family.'[7] Patricia Cohen has charted the uncertainty and ambivalence surrounding privacy, which was not then defined as 'the right to live as one wishes' or 'intertwined with personal freedom.' By mid-century, however, privacy, was becoming 'a precondition to marital intimacy.'[8]

It is precisely because servants changed positions often that they did not become so inured to the family sensory landscape in their imagination, and they compared different experiences when prompted in their memories. Their position gave them a certain distance in that intimate environment. Indeed there are two principal narrative threads which characterize their memory of the experience: they were either treated 'like one of the family' if they liked their employers or 'treated like a servant' if their experience was negative. Obviously these emotions shaped their recall, encapsulated by Esther Davis' account of her experience in country NSW during 1915 where the employer hid the food, locked the pantry and gave out meagre rations. Esther left after two weeks: 'I couldn't get anything to eat and they just treated me as a servant so I wanted to be treated as one of the family so I didn't stay.'[9] And in the latter accounts there is often a sense of grievance and injustice sustained over a long period that infuses their narrative authority.

If treated 'like a servant,' there was a multitude of ways that any physical contact or intimacy was restricted through physical and symbolic boundaries in the home. However, servants recounting their memories in the 1980s rarely saw themselves as 'victims,' and they presented as women with choices who took action when necessary, even if in fact their agency was circumscribed by a system which required them to work for specified periods of time or inform authorities if they left a place of employment. Moreover, the traditional 'victimhood' narrative of working in service, as Lucy Delap notes, fails to explain the many ambivalent or even positive accounts of their experiences which women narrate.[10] They might have referred to instances of ill treatment, but they also spoke of warm relationships, enjoyment of work and subtle means of negotiating the power relations. If all else failed, the act of leaving, changing jobs, became one of 'voting with their feet.' However, in relation to Indigenous servants, even the capacity to leave was severely limited, so they described other means of coping. Pauline Gordon, a Bundjalung woman from the Cootamundra Girls Home in NSW working in various parts of Sydney, was well aware of the constraints imposed by both the state and her employers. 'You couldn't go to anyone,' she said, 'and you couldn't say nothing and you couldn't throw a fit or you couldn't throw a tantrum or anything because you was nothing.' But, in her account, personal integrity was upheld by silence and deliberately distancing herself from her employers, often described as 'they,' who became all white people with power over her life: 'I never stepped over that mark to make myself familiar or anything like that . . . and I didn't want anything because they had nothing to give me.'[11]

The considerable literature in the field of sensory studies, which has now built up over the last 30 years, particularly historical research, has almost all been carried out on or in the variously defined 'public sphere.' It is possible that cultural historians have been drawn to sources which explore changes to social life for the largest number of people through the sensorium; and certainly the emergence of 'modern' life is primarily interpreted through the senses as a public phenomenon and indeed constitutes new understandings of the spatial.[12] No doubt personal memory sources present considerable challenges to the historian for exploring everyday life in the home. French historian Michelle Perrot cautions that that these sources 'do not constitute true documents of private life. Their contents are dictated by rules of propriety and a need for self-dramatization.'[13] However, arguably, all sources available to the historian are mediated in some way, and the British historian Carolyn Steedman's injunction to 'think with servants' has encouraged historians to work with this pervasive institution in different ways.[14]

Previous histories of domestic service in Australia have taken a broadly economic and social perspective, charting its decline after 1910.[15] It is certainly clear that the distinctions between servants and family, which were clearly demarcated through a household's spatial organization and social rituals keeping mistress and maid at a distance from each other, were gradually breaking down and the one- or two-servant household was a more common pattern from the 1920s. The work of scholars such as Victoria Haskins on Aboriginal servants and white employers, interpreted through a post-colonial lens, has made an important contribution to our understanding of the cross-cultural subtleties and nuances of this racial 'contact zone' in the home.[16]

The perspective charted here uses the sensory framework to focus primarily on intersectional class and gender power relations in the home. It draws on recent research by a number of English feminist scholars of twentieth century domestic service in England; namely, Alison Light, Judy Giles, Lucy Delap and Selina Todd.[17] All of these scholars argue the case 'to bring servants into history,' for the cultural and social significance of domestic service as central to the history of work as well as to the relations between women.[18] In an implicit comparison with the traditional labour history focus on masculine work and relations, Alison Light speaks of the 'schism between women which separated them in the very heart of "private life" and continued to produce tensions and structures of feeling formed as much in domestic service as the workplace.'[19] With obvious modifications for a racially diverse post-colonial Australia, and the exigencies of a very different environmental landscape that shaped work practices, I argue here that this research can inform the Australian context. This is especially the case because domestic service was a transnational occupation, and there was a considerable degree of movement between Britain and Australia in the sources of my study.

The Memory Archive

For the Anglo-origin population who grew to adulthood in the first half of the twentieth century, domestic service is now almost beyond living memory, except for the intergenerational recall of growing up with servants—it is a liminal, or threshold memory almost beyond the 'floating gap' past three generations. This is also largely the case with Indigenous Australians whose narratives are almost entirely of working as a servant, rather than employing them or growing up with servants in the house, although they often worked later into the twentieth century: until the 1960s in some regions and cities. There was also state-funded transnational migration of women from Britain to Australia before World War II, which was broadened to include Greek and Czech women from Europe in the 1950s.

However, through the emergence of sound technologies and written accounts, we have a somewhat dispersed account of working in that period in many archives across Australia. From 1986 to 1988 I carried out an extensive correspondence with nearly 200 women and conducted over 40 interviews with those who had worked in domestic service from the early part of the century. There are also repositories in state libraries which hold collections done by other interviewers all around Australia featuring this previously 'invisible' occupation which was the experience of so many women, both as employers and employees, at the time. Listening to them feels like pulling back time from oblivion, for even in the 1980s women were remembering some 50 and 60 years before.

Moreover, they were remembering their young selves before the intervening years of raising the school leaving age to 15 in the 1940s and 16–17 thereafter, and they often remarked on their own vulnerability or treatment, the heavy volume of work and the range of tasks at the age of 14 or 15. There are some interesting questions about how we can now 'remember' a period that is beyond our living memory, once removed through the technology that is not the subject of this essay, especially the role of oral historians who intervene in the process by which memories are transmitted or not (forgotten) across generations. Many now assume domestic service was an institution particular to class-based and hierarchical societies like Britain, especially since television series such as 'Upstairs, Downstairs' and 'Downton Abbey,' so pervasive is the media representation of the small percentage of servants who worked in the large estates.[20] Twenty-first century Australians mark themselves out from these experiences with the ideological assumption of a more egalitarian history that has been able to obscure the centrality of gender and class relations between women. And employers, not servants, often mentioned to the interviewer that 'everyone had servants in those days' to distance themselves from an exploitative system and separate the past from the present.

There is also a historiographical assumption that the numbers of women in domestic service began rapidly declining from the early twentieth century as women fled to the factories and other occupations when possible.[21] Research indicates that this argument needs some modification. The situation varied considerably from region to region and especially from rural areas to the urban. It is generally acknowledged that as general unemployment rose during the Depression years, the numbers of women in service increased. But it is also the case that, as many have noted, domestic service was by this time a transition job from the home to the workplace that young women took up from the age of 13 or 14 either because their families could not afford to keep them or because parents though of it as a 'safe' job for a vulnerable working class girl, or since their mothers had worked in service before them. Because domestic service was casual and not part of the employment figures, and women moved in and out of it and sometimes worked for only weeks or months before moving on, it is almost impossible to chart accurate figures for its persistence. The historian Carolyn Steedman describes British servants of the eighteenth century as 'culturally noisy but demographically elusive,' and this can be equally applied to those under study two centuries and in another country later.[22]

In the period under study then, domestic servants were almost all young women under the age of 25, working before they were married or transferred to another occupation (such as nursing; or cleaning in less constrained environments such as hospitals, hotels or boarding houses). In the 1980s they were rarely asked specific questions about sensory responses and analysis of their memories; however, over 30 years later, using this framework helps to illuminate very different aspects of the experience that might otherwise have been overlooked. For example, there is no clearer example of the dissonance in a new sensory landscape than when Doris Garard, who had lived in an orphanage, came from England in 1928 through Dr Barnardo's organization and began her first job, 'not quite' aged 14, on a sheep station:

> On arrival at the homestead the mistress told me to fill a bucket of water from the tank outside. There was no water laid on, also no electric; logs were the means of heating and cooking. I went out to fill the bucket, it was dark; as a child I was afraid of the dark. When the water started to fill the bucket it made such a noise. I looked around to see many eyes looking at me, I screamed. The mistress ran to see what was wrong. I pointed to the eyes and shouted 'wolves!' She clapped her hands and they all ran off. She said not to be so silly, they were sheep.

Principally through sound and sight, this small story encapsulates Doris's youthful fear and sense of disorientation, her lack of familiarity in a new country, and it remained forcefully in her memory, even though she returned to England within ten years.[23]

Intersensoriality

Sensory scholars such as Steven Connor remind us that it is rare to experience only one sense at a time and that the senses are 'inherently relational.' These 'synaesthesia and minglings,' he argues, are particularly evident in the sight–sound relation, so that 'the evidence of sight often acts to interpret, fix, limit and complete the evidence of sound.'[24]

Many of the references to seeing and hearing in the interviews or letters of the women who worked in service related to the impingement of the external world. Particularly in rural areas, there was flood and fire, and on a sheep station in Mungindi, NSW, the curse of the dust storms: '2 o'clock in the afternoon and it looked as if it were midnight, it was that dark,' said Jean. Despite closing all the doors and windows and stuffing up cracks with towels, the dust would still take days and weeks to clean up. Nor was there peace for long. The rains were followed by the grasshopper plagues: 'They came and it was like a big dark cloud in the distance, then you heard the noise.' These events were narrated as an indication of hardships and fear, but without rancour, as acts of nature.[25]

However, the enemy within, the sight and sound of rats, exceeded the limits of Clara Sadler's sensory tolerance: she had had migrated from England and in 1917 was working in service for the Flavell family at Kirribilli, an inner Sydney suburb. Seventy years later, she remembered the rats with disgust and fear. They left droppings and tooth marks, gnawing continuously, and she constantly imagined seeing them 'out of the corner of her eye,' but nights were worst:

> I knew they were in the house at night. We could hear them squealing and the boss used to set traps and then he'd drown them when he got them in a trap, he'd put the trap in the water till they drowned and then take them out. . . . And they used to come up and down the stairs and even eat holes in the stairways where we slept. I left there because I got scared. I hate rats.[26]

Sydney's 1900 outbreak of bubonic plague, carried by rats, had resulted in a slum eradication campaign but because Kirribilli was a suburb close to the water and docks, it was likely that the rat population continued to flourish in the uneven sanitary conditions.[27] Sadler's memory speaks of this fear of invasion as well as an odd mixture of disgust and relish about the method of killing them.

The sight–sound relation in domestic service is most evident in three features intersecting class and gendered rituals—mode of address, wearing uniforms and eating in the kitchen—that were utilized by employers to create and enforce social and cultural distance from their employees, to resist the intimacy that being 'one of the family' implied.

The simple act of speaking to employers could be a fraught experience for women employed to work in a household. In Carmen Denning's interview for a job at Bellevue Hill, Sydney, during the 1920s, Mrs Sterling declared that 'Carmen' was too good a name for a servant and she would name her 'Connie.' As a way of resisting, Carmen didn't respond often when Mrs Sterling called. But in this home, there was a politics to her renaming as well. Mr Sterling asked what her name was: 'It is Carmen but Mrs Sterling wants to call me Connie.' Carmen then remarked: 'I then heard him say to his wife through the swinging door in the dining room "you had no right to call her that." ' In this case, the husband was overheard to intervene on the servant's behalf, mediating the relationship between women, but this was not common, and many recall that the male employer barely spoke to the maid at all.

Carmen also referred to several instances where servants were 'dehumanized' through the speech and behaviour of their employers in Sydney during the 1920s. Her next employer, Mrs O'Brien, avoided speaking to her directly and called her as the 'domestic' or the 'servant.' In a larger household with a number of servants, Mrs Toohey in Potts Point gave orders for the day in her corset, showing no apparent self-consciousness. 'To them you weren't real,' says Carmen—'servants were sexless.' These instances obviously affected Carmen, who was gently reproving of her treatment during interviews.[28]

Similarly, Noreen Shirt described as a 'strange thing' that neither her mother and aunt, who were domestics, nor herself, who also worked in service during 1936, were 'allowed to use their given names.' Her mother had to answer to Ellie and her aunt was known as 'Nurse,' and when Noreen went to work as a general maid at Rose Bay in Sydney, 'my employer Mrs Stevens wanted me to answer to Jane as she thought my given name was "unsuitable." ' However, 'I politely requested her to call me by my name and she eventually got used to it,' adding that 'very little changed in the years between my mother's service and my own.'[29] Arguably, Noreen Shirt's resistance to this practice was in fact part of changing domestic relations and a stronger sense of an emerging self.

Many maids spoke with dislike about being required to speak indirectly about employers as 'master or mistress' and in direct speech 'as Sir or Madam,'[30] and even more resented having to address the children by the same moniker. When servants mentioned speaking with employers in interviews, it was always with this resentment, since these modes of speech underlined their inferior position. They often charged their employers with 'snobbery,' a pejorative term referring to pretension or affectation of class superiority. For example, Mary Tweedy, working in Melbourne for a member of the legislative assembly, Mr Linton, and his wife, called them 'snobs' since she always had to answer the telephone by saying, 'The residence of Mr Linton MLA.'[31]

Many women who worked in service also described their uniforms in great detail in the 1980s in correspondence and interviews, although they were more positive about the wearing of these than about the modes of speech required of servants, perhaps because uniforms were as often designed for function as for performance. Moreover, many changed uniforms in the middle of the day, which was a common practice because families entertained in the afternoon and evening, so the servant had to be *seen* in front of visitors to the home, or 'putting on the dog' as one disparagingly called it. But other youthful girls loved the novelty of the clothes and being observed by others. Mary Tweedy had her first job in England, where she was employed as nanny at age 14 to a child of 20 months, and she often took the child for a walk:

> I had blue uniforms and had great trouble managing the stiffened collar and cuffs—but I thought I was Christmas with my navy blue uniform overcoat and navy hat, rather like a squashed 'pork-pie' style of men's felt hat.[32]

Both the time of wearing and the nature of uniforms varied, especially in rural areas where the manual work was so heavy, but almost no employees ate with the family. They ate in the kitchen alone or with other servants.

One child of an employer remembers thinking that servants never ate at all because they didn't see them doing so while they were growing up.[33] This seemed to be the one factor across the period whereby distance between maid and employer was not breached, and some found it very lonely while others preferred to eat this way. Irrespective of where they ate, women often spoke of being treated 'like one of the family,' so eating in the kitchen away from the family may have been seen as less stressful since it was 'private' or not under the watchful eye of the employer. Many years later, Mrs I. E. Olde remembers that when she first left school in 1935, she worked in residential domestic service for a doctor in a small NSW country town of about 900 people. She writes a vivid multisensorial description of the process of work in relation to her female employer:

> When serving her midday meal I had to wheel her food . . . across a passageway, lifting the traymobile down one step and then up another, first of all donning a little white apron then go back to the kitchen, and wait until a little silver bell was rung for me to go in and receive my meal, which (thank goodness) I could eat in the kitchen while seeing to the dessert. Then came the same procedure: Go in, clear the table, wheel dirty dishes out, reset the traymobile . . . when there were visitors I found this very humiliating. After all I was not yet 15! Very nerve-wracking too.[34]

Mrs Olde uses her young age here as the vehicle to speak of the remembered injustice and inequality, and a particular indignity was reserved for others from outside the space of the home seeing or watching her do the work.

Touch

Rarely mentioned in interviews, touch is a proximal sense that remains historically muted in studies of the public sphere but is central to the memories of servants in the homes where they worked. On some occasions, servants referred to 'the beautiful things' in a household, which they held, observed, dusted and cleaned, and could describe the feel of them and their colours with accuracy. However, they rarely referred to touch between people in the household except in shared moments of great intimacy or memories of attempted violation of their own bodies and space. Hilda was an assisted migrant from England as a child with her family in 1913, but moved between England and Australia with her work as a servant. In 1918 she looked after an old women, a Mrs Davey, who required assistance to bathe from another servant as well as herself. She said they would 'stifle giggles when they pushed her breasts and body into stays and laced them up the back.' However, 'As a special treat, the old lady would love to be taken to a warm and sunny spot in the garden and have her hair attended to. Hilda would remove the pins and brush her hair gently for about half an hour, it made Mrs Davey's day.'[35] While the nature of this intimate touch reinforced social distinction and older female authority, the description also attests to kindness and friendship. Similarly, in 1911, Agnes Rowe of Queensland remembers growing up with the South Sea Islander women who remained behind as servants after the men were sent back to the Solomon Islands. 'They were the kindest of people,' she wrote, 'those black hands used to bathe us, put us to bed. Any move or cry in the night brought her in. A candle in her hands. "Don't cry to disturb your mother—come with me." ' Again, there is a visual memory of skin colour intermingled with the sense of touch from a child's point of view.[36]

The historian Elizabeth Harvey reminds us that touch, 'more than any other sense,' as well as being the most dispersed throughout the body, 'establishes our sentient border with the world.'[37] But touch was also central to gender relations of course, and is associated with sexuality. The most consistent form of touching that has remained in the memory of women in service, and the principal means of charting its limits and force, is through what we now call sexual harassment. 'Men of any rank,' says the historian Mark Smith, 'saw the female body as always open to touch and therefore possession.'[38] But, Smith argues, by the nineteenth century, touch became an important signifier of class differences between manual and non-manual workers. Central to new norms was personal discipline, between rough and refined touch, and establishing, says Smith, a 'temporal gap between desire and its gratification—the ability to resist direct contact, to pause and keep a distance versus the submission to tactile impulse, the grasp, the clumsy lunge, the heavy-handedness of handling.'[39] In the case of domestic service, this was a matter of young women evading the advances of men in the household, but their response was always constrained by the inability to speak about it to relevant authorities, parents, female employers or guardians, and to be questioned if they did so. So, Joan Rissman's first employers after she left school at 14 were a German couple in outback Queensland. 'It wasn't too bad at first,' she explained; the woman employing her was 'very nice,' but: 'We never had sex explained to us by [our] step-mother. For some reason in those days it was never mentioned which was a shame' because:

> After I had been in that job for a while, the woman's husband came downstairs one morning, as I was trying to light the fire, and started rubbing his hands up and down my legs and my senses told me that wasn't right, but before he went any further, he heard his wife coming downstairs and he walked away. I felt so self-conscious that I thought she may have known what was going on. So I left not long after that.[40]

It is clear from this narrative that in the relatively privatized space of the home, men waited for their opportunities, for the moments of being completely free from observation (just as they did later in the privacy of the boss' office for the secretary, or in the rooms of live-in colleges), and that the young Rissman was made to feel a sense of complicity: 'feeling dirty' is how it was often expressed.[41] However, while scholars such as Corbin and Smith refer to the importance of 'the delicacy of the hand' in marking out class differences, it was clear from servant experiences that this was not a factor in male predatory behaviour. Instead, the hand came to symbolize the male touch they abhorred and it could become the focal point of their fear. This was particularly the case with women who were sent out from orphanages or state institutions. In this case of an Aboriginal woman sent out from Cootamundra Girls Home, her anger echoes down the years. Her first job in rural NSW on a farming property underlined her isolation and vulnerability:

> He [employer] used to try and maul me. At first I thought, lovely old man making me feel welcome. . . . I really thought he was interested in my welfare, but it was more than my welfare he was interested in. He soon showed that after a couple of weeks—wandering hands—I nearly broke his bloody hands . . . I used to be that scared when I went to bed at night. . . . I'd sneak knives out of the kitchen and I was scared he'd come and attack me.

For this young woman, who is not named in the text, the matron at Cootamundra, whom she rang about the harassment, did not believe her story, but she ran away back to the home anyway.[42]

'Wandering hands' or the unwanted touch of the male hand was the experience of many women, and for some it became the signature of their understanding of men in general. 'They were all like this,' said Kit and Alma Hughes, who had come out from England in 1923, to the Western Australian interviewer Sally Kennedy. In one of Kit's first jobs at 14:

> I went sort of away in the country as a ladies companion. The lady was the daughter of the house and there was the old father and the son and the old man used to smack me on the bottom every time I walked past him and make a pass at me and he had palsy!

She was terrified 'to go in the room when he was there on his own or walk past him on the verandah.' Alma Hughes was also 'too afraid to tell parents. In those days you didn't discuss those things with your parents.' Later, at a dairy farm, it happened with another employer who 'got me, shut me, got me shut in the freezer. They get you in ice in the freezers, you get shut in the freezers and make passes at you and you couldn't get out . . . we were fairly innocent in those days and we didn't know much.' In her memory of being a young girl, they were 'huge men' pushing her into spaces where they could be unobserved. The Hughes sisters' solution, like those of others, was to move on from job to job.[43]

But some women did speak to others about their experiences. In 1931, when she was 16, Mae Elliot was in service on a rural property 16 kilometres away from her family in Victoria, because they desperately needed the money during the Depression, and she was being harassed by the young men of the family. She was unhappy and felt the conflicting pressure of the need to contribute to the family income:

> The boys used to try and peep through the blind at me getting undressed or would watch me go into the lavatory which was right away from the house. The eldest boy was always trying to kiss and fondle me and I hated it. I was there two years and when my father heard about the boys trying to peep at me he said I must leave. He was a very strict father.[44]

In this instance there were clearly very few spaces in the household for Mae, or indeed other women, to be private or 'unobserved,' especially if they shared a room for sleeping or a bed was made up for them on a verandah.

Of course the manual work of servants required constantly touching objects and parts of the house, and many referred to 'having to get down on their hands and knees' to wash floors, which could be painful until the body became used to it. Indigenous woman Eunice Robinson, who told her interviewer that she liked working, grew up in and around Tabulam in northern NSW. She started 'doing for' a woman with five adopted children who was 'allergic to dust': 'I had to get down on my hands and knees, scrub, polish and then look after these kids as well, it was hard. Look, both knees was all big boils—housemaid knees they call it. But when you get over that you don't get them no more.'[45] As Eileen Boris contends, 'the body marks class in material and symbolic ways.'[46]

Direct and Implied Smell

Consideration of smell in the home raises the important issue, as the cultural historian Alain Corbin notes, 'of the frontier between the perceived and the unperceived, and even more of the norms which decree what is spoken and what left unspoken. We need, in fact, to be careful not to confuse what is not said with what is not experienced.'[47] There is no doubt that the first half of the twentieth century environment in the home was pervaded by smells, but there is little mention of it in the memories of women who worked in service. This is particularly the case, for example, with the daily task of emptying night chamber pots. A great many women in my study refer to this task, universally disliked, but they never mention the smell of them. It may be, as Corbin says again, that 'the banal is frequently silent' or that it was considered inappropriate to discuss it even in the 1980s, or that the task itself connoted one of the most demeaning aspects of the position and the smell was assumed. But even alluring kitchen smells of food went unremarked over the course of this project, so it may be related to the ubiquity of different smells that were forgotten. Instead, the unusual was often remembered. Elsa Shillingford, for instance, was employed at 19 or 20 in 1929 in a soldier settlement dairy farm on the way to Tarcutta, in southwest NSW. She narrated a story involving bedbugs with a sense of disgust and horror: she described at first finding one of the babies of the household covered with bugs in the night when she went to comfort her crying. I asked, as the interviewer indicating my ignorance, 'what kind of bugs were they?' and she replied in a testy tone:

> There is only one sort of bugs my love and they are bed bugs and once you have seen one or smelt one there is no other smell. It's just yuk. It's just awful. We got the beds out and he [employer] wanted to know why I wanted pounds of alum. Well we had nothing.[48]

In a frenzied way, Elsa poured boiling water and alum all through the beds and in the holes in the wooden furniture. Alum has a powerful antiseptic quality for killing germs, but it did not seem to be particularly effective here. Bedbugs are described by scientists as having 'an offensive, sweet, musty odour from their scent glands, which may be detected when infestations are severe.'[49]

In the enforced intimacy of the home, the smell of other people, their bodies, was also rarely mentioned, but everyday smells such as these may not have remained in servants' memories. Employers thought it important for servants to be separated from family sleeping arrangements, and this was symbolic of the limits to intimacy, except when families could not afford such distancing, as was often the case in a smaller house or urban apartments.

But servants not only moved through a landscape, experiencing smells, they also created odours. Elsa Shillingford related an incident that refers indirectly to her own body odour:

> In my day you never bathed when you had your period which of course is the time you needed the bath. And of course bathing was a thing that everybody knew that it was the day that you had your bath. And I didn't get a bath and she said: 'Why aren't you getting your bath ready?' I said 'Oh no, I can't; I've got my period' and she buggered it up. She said 'No. You definitely are going to have a bath' and I mean I never look back after that when I'd get it.[50]

This story tells us not only of issues related to personal hygiene but also a trusting *in loco parentis* relationship between servant and employer. While none of the women who worked in service spoke about their own body odour, both the children growing up with servants and employers remarked on the smell of sweat and other smells associated with particular servants.

Conclusion

Domestic service in the home for the first half of the twentieth century both constituted intersecting class and gender relations and was the site of their increasing contestation. These many and varied sensory memories of working in domestic service from the point of view of the servants themselves have provided an unusual insight into the workings of the household in Australia (and to a lesser extent England) as a place of complex negotiations over privacy and personal space. They reveal that the home was a place where commonplace work and daily rituals were carried out, and that it was also not a 'safe' place for young women but an environment in which they needed to be constantly alert sensorially.

Some aspects of these women's experiences, the work and conditions, had resonances with what we know about the long history of servant-keeping. All manner of cultural representations, for example, have shown us that men have been harassing, seducing and raping servant girls for centuries (Mozart's opera *Don Giovanni* comes immediately to mind); and the work process of manual labour changed little before electrified appliances. (Using a broom to sweep the floor for instance, has a very long history.) However, the focus on the sensory, separately and intermingled, adds an important layer to the history of domestic life. First, it alerts us to the ambivalences by both employers and workers about intimacy, and the shifting boundaries and understandings of privacy during this period. Second, this study gives us a more capacious understanding of what being in service meant for young women, the construction of individual subjectivities and sexual difference. As well, it allows us to explore the meaning of their remembrance across the years, and the force of a double mediation of memory sources: reliving experience through letter and interview in the 1980s, with the historian's intervention to interpret the meaning of it 30 years later.[51]

Finally, the sensory also proves to be remarkably evocative of the experience, to bring it into memory with imagination through the personal sources in a way that is usually left to novelists and film-makers to re-create. It brings the previously unexplored world of private life to sensory studies for historians.

Notes

1 Photograph taken by Paula Hamilton, 15 January 2015.
2 Paula Hamilton interview with JM Balmain, Sydney, 15 January 2015.
3 Kate Lacey includes here a useful discussion of the public/private binary and its limitations. See Lacey 2013, p. 6.
4 Rosenfeld 2011, p. 318.
5 Mrs J. E. Rissman, Toowoomba, Queensland: letter to Paula Hamilton, 29 May 1986. (Domestic service archive in author's possession.)
6 Margaret Mackie, letter to Paula Hamilton, 28 July 1986. Electrified bells were installed during the 1920s.
7 Giles 2004, p. 22. See also Flanders 2014 and Schwartz 2012.

8 Cohen 2013, p. 183.
9 Paula Hamilton interview with Esther Davis, 20 February 1987, NSW Bicentennial Oral History Project, transcript and tape in SLNSW, MLOH 5163.
10 Delap 2011a, p. 28.
11 Inge Riebe interview with Pauline Gordon, 3 March 1988, NSW Bicentennial Oral History Project, transcript and tape in SLNSW MLOH 5163. Gordon and her family were from the north coast of NSW, near Grafton, and her father was an Aboriginal rights activist, so she had a stronger sensitivity to racism and power relations.
12 See for example Smith 2004; Morat 2014, particularly the work of Emily Thompson.
13 Perrot 1990, pp. 3–4.
14 Steedman 2009, pp. 13–14.
15 Higman 2002.
16 Haskins 2013. See also Haskins and Lowrie 2015.
17 Light 1991; Giles 2004; Todd 2009; Delap 2011a and 2011b. See also Howes and Classen 2014, pp. 65–92 and Classen 2012, p. 104.
18 Giles 2004, pp. 11–12; Delap 2011a, pp. 2–25.
19 Light 1991, p. 219. See also Todd 2014 *passim*.
20 Toynbee 2014.
21 Higman 2002, Chapter 2.
22 Steedman 2007, p. 14.
23 Letters from Doris Garard, Essex, to Paula Hamilton, 28 April and 11 June 1986.
24 Connor 2004, pp. 153–54.
25 Letter from Jean Mullier to Paula Hamilton, 8 January 1987.
26 Paula Hamilton interview with Clara Sadler, 21 September 1987. NSW Bicentennial Oral History Project, MLOH.
27 See Ashton 1985.
28 Paula Hamilton interview with Carmen Denning, Sans Souci, 21 November 1986. In author's possession.
29 Noreen Shirt, letter to Paula Hamilton, 23 May 1986.
30 Barbara Harper, Clifton Beach, Tasmania, letter to Paula Hamilton, 25 May 1986.
31 Mrs D. Gyles, Queensland, letter to Paula Hamilton, 18 May 1986.
32 Mrs D. Gyles, ibid.
33 For example, letter from Mrs J. V. Lockrey, Dubbo, NSW, 21 February, 1987; letter from M. Doubleday, 14 April 1986; letter from Emily Steel (Rossell), Orange NSW, 9 April 1987.
34 Letter from Mrs I. E. Olde to Paula Hamilton, 14 April 1986.
35 Letter from Klaenna Pearson, Werribee, Victoria, to Paula Hamilton, 29 June 1986. 'Hilda' was her mother and she is telling her story in the third person.
36 Letter from Agnes Rowe, Queensland, to Paula Hamilton, 10 March 1987.
37 Harvey 2011.
38 Smith 2007, p. 101.
39 Ibid., p. 108.
40 Joan Rissman, Toowoomba, Queensland, to Paula Hamilton, 29 May 1986.
41 See Paterson 2007, p. 164 and Robbins 1993. Merleau-Ponty notes that to touch is always to be touched as well. Cited in Harvey 2011, p. 387.
42 Extract from interview with no name in Kabaila 2012, p. 42. See also Haskins' (2013) exploration of a late nineteenth century case.
43 Sally Kennedy interview with Kit and Alma Hughes, April 1977. Battye Library of Western Australia Oral History Collection, OH209 A/r and T/r, pp. 4–6.
44 Letters from Mae Eliot to Paula Hamilton, 7 June and July 1986.

45 Inge Riebe interview with Eunice Robinson, 29 February 1988, NSW Bicentennial Oral History Project. MLOH 5163.
46 Boris 2013, p. 81.
47 Corbin 2005, p. 135 and see also Corbin 1986.
48 Paula Hamilton interview with Elsa Shillingford, NSW Bicentennial Oral History Project, MLOH 5163.
49 See Drobnick 2006 and Department of Medical Entomology, University of Sydney: http://medent.usyd.edu.au/fact/bedbugs.html.
50 Paula Hamilton interview with Elsa Shillingford.
51 As Judy Giles (2004, p. 24) argues, 'if we are to understand the past as something more than a backwards projection of the present, it is essential that we attempt to explore the ways in which subjectivities, memories and imagination were constituted differently—differently *then* from *now*.'

References

Ashton, P., 1985. 'Rats: Public health in Sydney, 1900', in D. Stewart, ed., *Case studies in Australian history*, Melbourne: Heinemann Education, pp. 94–109.
Boris, E., 2013. 'Class returns', *Journal of Women's History*, Vol. 25, No. 4, Winter, pp. 74–87.
Classen, C., 2012. *The deepest sense: A cultural history of touch*, Champaign, IL: University of Illinois Press.
Cohen, P., 2013. *Family secrets: The things we tried to hide*, Viking; republished 2014, London: Penguin.
Connor, S., 2004. 'Edison's teeth: Touching hearing', in V. Erlmann, ed., *Hearing cultures: Essays on sound, listening and modernity*, Oxford: Berg, pp. 153–72.
Corbin, A., 1986. *The foul and the fragrant*, Cambridge: Harvard University Press.
Corbin, A., 2005. 'Charting the cultural history of the senses', in D. Howes, ed., *Empire of the senses: The sensual culture reader*, Oxford & New York: Berg.
Delap, L., 2011a. *Knowing their place: Domestic service in twentieth-century Britain*, Oxford: Oxford University Press.
Delap, L., 2011b. 'Housework, housewives and domestic workers', *Home Cultures: The Journal of Architecture, Design and Domestic Space*, Vol. 8, No. 2, pp. 189–209.
Department of Medical Entomology, University of Sydney, 'Bed Bugs', available at http://medent.usyd.edu.au/fact/bedbugs.html. Accessed 28 January 2015.
Drobnick, J., ed., 2006. *The smell culture reader*, Oxford & New York: Berg.
Flanders, J., 2014. *The making of home*, London: Atlantic Books.
Giles, J., 2004. *The parlour and the suburb: Domestic identities, class, femininity and modernity*, Oxford & New York: Berg.
Harvey, E.D., 2011. 'The portal of touch', AHR Forum, *American Historical Review*, Vol. 116, No. 2, pp. 385–400.
Haskins, V., 2013. ' "Down in the gully and just outside the garden walk": White women and sexual abuse of Aboriginal women on a colonial Australian frontier', *History Australia*, Vol. 10, No. 1, April, pp. 11–34.
Haskins, V.K. and Lowrie, C., eds., 2015. *Colonization and domestic service: Historical and contemporary perspectives*, New York: Routledge.
Higman, B., 2002. *Domestic service in Australia*, Melbourne: Melbourne University Press.
Howes, D. and Classen, C., 2014. *Ways of sensing: Understanding the senses in society*, London & New York: Routledge.
Kabaila, P., 2012. *Home girls: Cootamundra Aboriginal home girls tell their stories*, Canberra: Canprint Publishing.
Lacey, K., 2013. *Listening publics: The politics and experience of listening in the media age*, Cambridge: Polity Press.

Light, A., 1991. *Forever England: Femininity, literature and conservatism between the wars*, London: Routledge.

Morat, D., ed., 2014. *Sounds of modern history: Auditory cultures in 19th and 20th century Europe*, New York & Oxford: Berghahn Books.

Paterson, M., 2007. *The senses of touch: Haptics, affects and technologies*, Oxford: Berg Publishers.

Perrot, M., 1990. 'Introduction', in P. Aries and G. Duby, eds., *A history of private life, Vol IV: From the fires of revolution to the Great War*, Cambridge, MA: Belknap Press, pp. 3–4.

Robbins, B., 1993. *The servant's hand: English fiction from below*, Durham & London: Duke University Press.

Rosenfeld, S., 2011. 'On being heard: A case for paying attention to the historical ear', AHR Forum, *American Historical Review*, Vol. 116, No. 2, pp. 316–34.

Schwartz, L., 2012. 'Rediscovering the workplace' (Review), *History Workshop Journal*, Vol. 74, Autumn, pp. 270–77.

Smith, M., ed., 2004. *Hearing history: A reader*, Athens, GA: University of Georgia Press.

Smith, M.M., 2007. *Sensing the past: Seeing, hearing, smelling, tasting and touching in history*, Berkeley, CA: University of California Press.

Steedman, C., 2007. *Master and servant: Love and labour in the English industrial age*, Cambridge: Cambridge University Press.

Steedman, C., 2009. *Labours lost: Domestic service and the making of modern England*, Cambridge: Cambridge University Press.

Todd, S., 2009. 'Domestic service and class relations in Britain 1900–1950', *Past and Present*, Vol. 203, May, pp. 181–204.

Todd, S., 2014. 'Class, experience and Britain's twentieth century', *Social History*, Vol. 39, No. 4, pp. 489–508.

Toynbeld, P., 2014. 'What if Downton Abbey told the truth about Britain?' *The Guardian*, 23 December. Available at: http://www.theguardian.com/commentisfree/2014/dec/22/downton-abbey-truth-about-britain

13 Botanical Memory
Materiality, Affect and Western Australian Plant Life

John Charles Ryan

On a clear spring afternoon, I follow local conservationist Ian Smith around the rocky outcrops and through the wooded gullies of Mount Matilda in Western Australia (WA). After years of farming the region's flatlands, Ian dedicates his retirement to the promotion and protection of wildflowers—those endemic counterparts of canola and soy. Named the Wheatbelt, this 15,540 square kilometre part of the state lies between metropolitan Perth to the west and the arid goldfields to the east (see Figure 13.1 for a map of the region). Within this predominantly agricultural landscape, locales such as Mount Matilda are ecological islands, regarded by settlers as unsuitable for pastoral activities because of rugged terrain and noxious plants. Regularly on our ascent to the hill's highest point, Ian gestures at flowers with his walking staff—handcrafted from a local tree known as gimlet (*Eucalyptus salubris*)—as we exchange observations, ideas and recollections.

Like Ian's staff, the hill and its plant life bear testimony. All have survived the austerity of the environment and the depredations of settler history. The gracefully convoluted walking stick embodies and indeed facilitates Ian's attachment to this enduring sanctuary. 'The wood is so strong. I've used this staff for years, but there is not much wear. I've stained it with boot polish, and varnished it. That's my name there.'[1] Ian's gimlet staff further points to the material and affective dimensions of human memory, developed in this chapter through the concept of botanical memory. In previous work, I have defined botanical memory as a form of environmental or place-based memory focused on people's remembrances of plants, and on individual and collective practices that instigate and sustain such recollections.[2] In my field interviews with Ian and other floristically minded individuals, the interconnections between plant materiality—the spikiness, stickiness, smelliness of living plants, as well as the qualities of plant-based objects—and human memory are palpable.

In this chapter, I historicize and analyse ethnographic content, exploring a material-affective understanding of botanical remembrance as a 'dialogic encounter'[3] between people, plants and places. My interview with artist Holly Story reveals her engagement with the edible resin of marri (*Corymbia calophylla*) for producing works of art in which the endemic tree does the painting.[4] In another interview, the WA artist Nalda Searles explains

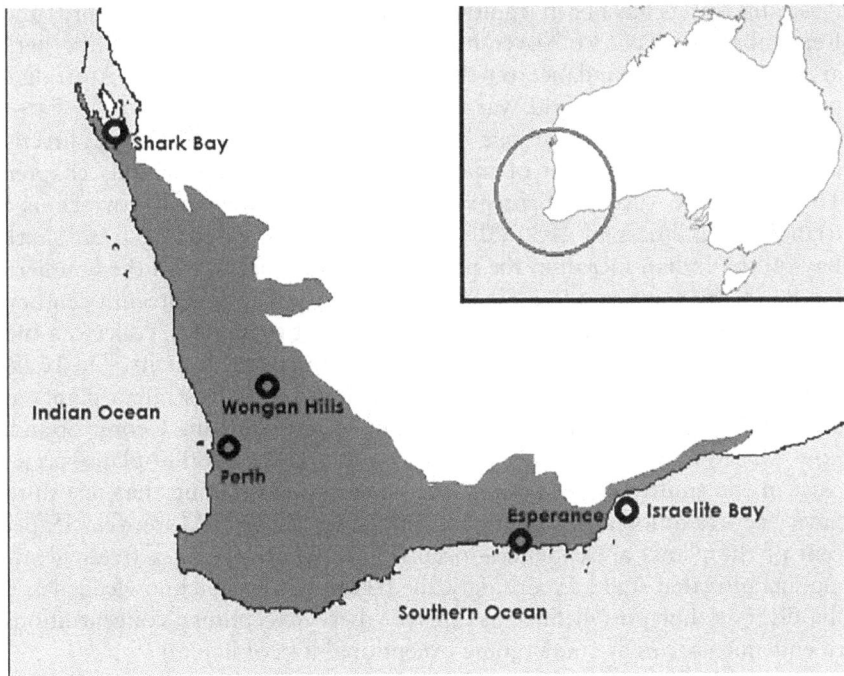

Figure 13.1 The Southwest Australia Ecoregion of Western Australia, extending from Shark Bay to Israelite Bay (shaded triangular area) (2013).

Credit: Gossipguy (Public Domain via Wikimedia Commons). (CC BY-SA 3.0)

that textiles of balga (*Xanthorrhoea preissii*) embody memories of the plant and her creative process.[5] In comparison, plant propagator Kevin Collins employs banksia (*Banksia* spp.) parts, particularly their flowers and seed cones, to catalyse memorable experiences for visitors while evoking his own sensory recollections of the small trees.[6]

Background to Botanical Memory: Place and Practice

For many cultures and eras, plants have served a crucial role in remembering. A salient example from the history of art is the English artist Dante Gabriel Rossetti's painting *Lamp of Memory* (1881) depicting Mnemosyne, the Greek personification of memory, remembrance and language, gazing forward hypnotically as she holds a golden beacon. At her altar, in front of her flowing green dress, there lies a small yellow pansy with a dark centre and a yew sprig nearby. Associated mythologically with the Greek god Eros, the pansy symbolizes Mnemosyne's transcendence of human thoughts, whereas the yew represents immortality.[7] Other notable examples include the herb rosemary that has long been associated with human memory. In Australia, sprigs have been traditionally worn on Anzac Day (25 April) and Remembrance Day (11 November), the cultural significance of the herb amplified by its abundance on the Gallipoli peninsula, where Australian troops landed during World War I.[8]

In addition to their symbolic importance, plants are also more directly involved in the formation of human memory, as argued in this chapter. The Southwest Australia Ecoregion (or Southwest Botanical Province) is a triangular landmass of 493,000 square kilometres, extending from Shark Bay on the Indian Ocean in the northwest to Israelite Bay on the Southern Ocean in the southeast (see Figure 13.1).[9] In the late seventeenth century, the region's botanical uniqueness began to attract naturalists, collectors and explorers, many of whom were astounded by the plant diversity.[10] In 1699, William Dampier made the first known collection of Australian plants at Shark Bay, followed in 1791 by Archibald Menzies at King George Sound, now Albany.[11] Modern botanists estimate that close to 8000 plant species exist in the southwest—a remarkable quantity, considering that one third have been taxonomically identified only since 1970.[12] Moreover, 35 per cent of the plants are endemic—occurring nowhere else in a freely growing, uncultivated state. Accordingly, the region has been acknowledged as a 'biodiversity hotspot,' defined as a place where 'exceptional concentrations of endemic species are undergoing exceptional loss of habitat.'[13]

Despite its historical renown, by the year 2000, the southwest's vegetation had been reduced to a meagre 10 per cent of its original 300,000 square kilometres.[14] Of critical importance to conservation of extant diversity is the botanically rich *kwongan*, an Aboriginal Nyoongar term denoting the unique sandplain vegetation.[15] Mount Matilda, the site of my perambulatory conversation with Ian, is ensconced there. Cleared extensively for agriculture and pasture, the *kwongan* is currently threatened by dryland salinity, plant disease, invasive species and habitat fragmentation.[16] Regarding the southwest, botanists Stephen Hopper and Paul Gioia assert that 'fundamental changes in attitudes toward land use and the intrinsic value of plant life are needed to go hand in hand with a commitment to protect, repair, and restore native vegetation in the face of uncertainty.'[17]

Following this entreaty to foster the appreciation of 'intrinsic value,' conservationists have created initiatives, such as the archival project FloraCultures, to promote the cultural heritage of the region's plant life.[18] One of the aims of FloraCultures is to highlight the intrinsic value of memory by conserving the plant-based recollections of individuals. Botanical memory contains rich cultural information about plants (for instance, collective community practices involving wildflowers), but also reveals critical ecological data, such as the former geographical distributions and flowering times of orchids. Simply put, memory is a vital element of southwest biocultural heritage. These projects conceptualize the heritage of southwest flora through the myriad ways in which plants inspire individuals and communities across

time, including through memory.[19] Nonetheless, whereas the scientific significance of the flora has been well-articulated and supported,[20] expressions of the impact of plants on WA culture in the past, present and future
remain dispersed across heterogeneous sources: in letters, journals, newspaper articles, blogs, websites, poetry, novels, plays, performances, paintings, photographs, sculptures, crafts—and in the memory-rich narratives of
activists, artists, conservationists, scientists, tourists and other plant enthusiasts. These stories constitute a vital, though relatively unprobed, facet of
the southwest flora's value—its 'intangible cultural heritage,' characterized
as 'forms of cultural heritage that lack physical manifestation [and evoke]
that which is untouchable, such as knowledge, memories and feelings.'[21]
Hence, remembering plants and the heritage (including human practices)
that surrounds them is critical to conservation; memory work can enhance
both cultural heritage and scientific conservation, an intersection explored
by Virginia Nazarea in her book *Cultural Memory and Biodiversity*.[22]

As the rates of floristic decline indicate, the viability of intangible plant-
based heritage is in jeopardy. Therefore, cultural conservation should be a
concern in the region, in conjunction with the biological conservation aims
of protecting living plants in their habitats. Referring to a global context,
UNESCO acknowledges that 'oral traditions and expressions . . . social
practices, rituals and festive events . . . knowledge and practices concerning nature' are intangible heritage priority areas.[23] Mindful of the value of
such heritage, I have interviewed individuals in the southwest since 2009 for
their memories of flora. Recently, I have begun to conduct film-based oral
histories focusing on the locales and species of resonance to the individuals,
but with an ongoing emphasis on the *kwongan*, which includes the rapidly changing metropolitan Perth area. My interviewees have had various
attachments to southwest plants; some were recently landed tourists who
had followed, in their caravans, the seasonal blossoming of wildflowers
from Shark Bay to Esperance, whereas others had life-long, site specific histories of physical, emotional and spiritual attachment to flora.[24] Where possible, the interviews occurred in 'the field'—for example, during wildflower
walks, on the grounds of a botanical garden or in the studio of an artist.
Such field settings enabled particular interactions between the interviewee,
myself as ethnographer and the plant-based prompts of seeds, leaves, flowers, roots, smells, tastes, sensations, artifacts, implements and artworks.

In relation to this ethnographic practice, a material-affective theory of
memory frames the bodily, sensory, emotional and terrene (*earthy*) features
of botanical recollection. In developing such a framework, I build upon the
work of American writer Henry David Thoreau (1817–62) and scholar C.
Nadia Seremetakis. Developed with reference to the American Northeast
flora, Thoreau's botanical memory is conspicuous in his posthumous works.[25]
In particular, *Wild Fruits* (2000) crystallizes the nodes between materiality
(of the plants themselves and related objects) and affect (of expressiveness,
emotion, embodiment, sensoriality and relationality). Thoreau conceived

of these writings as a 'Kalendar'—involving the documentation of the seasonal phenomena of Concord, Massachusetts, intermixed with historical, philosophical and experiential reflections. His journal was a substrate for 'material memory,'[26] hybridizing his personal recollections with those of other writers, both classical and contemporaneous. As a result, critics have characterized Thoreau's eco-literary corpus as 'a project in memory.'[27]

An entry from *Wild Fruits* presents a haptic recollection of *Desmodium*, a genus of plants in the pea family having the common name tick trefoil. The seedpod of *Desmodium* is known as a *loment* (see Figure 13.2). Each seed is enclosed in a triangular sheath that readily detaches from adjoining seeds and bears barbed hairs to enhance its dispersal by clinging to animals. Thoreau recounts walking with a friend through a thick patch of trefoil along the Concord River, the 'green scale-like seeds densely covering and greening our legs [and amounting] to a kind of coat of mail.'[28] With curiosity, humour and amity, Thoreau's rendering is attentive to the materiality of the loment, the two men's clothes and the ebbs and flows of walking: 'It was the event of our walk, and we were proud to wear this badge.'[29] His companion's fervent devotion to the Velcro-like seed badge even 'betrayed a certain religion about it' because when he 'reappeared for a walk a day or two after, his clothes were nearly as well covered as at first.'[30]

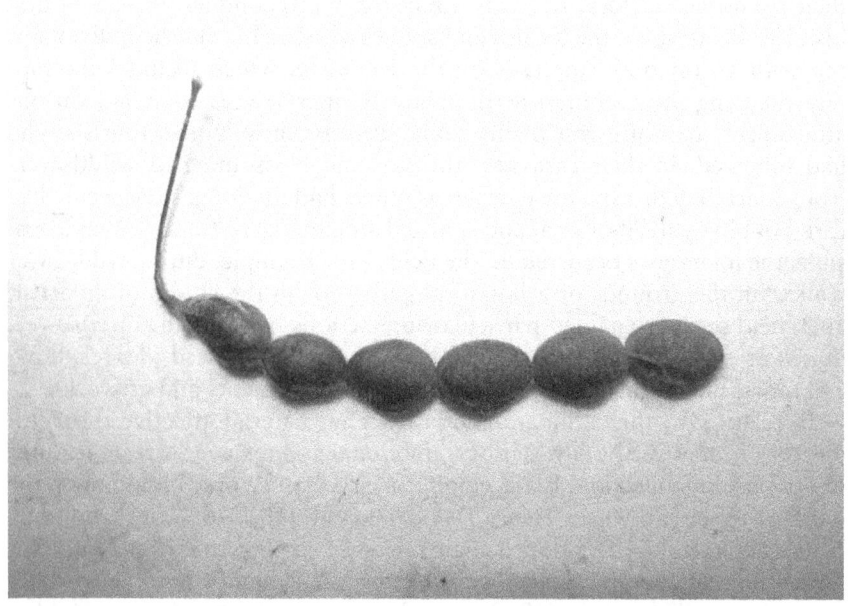

Figure 13.2 Loment (seedpods) of trefoil (*Desmodium* spp.), Hawaii (2003).
Photo: Forest and Kim Starr (Public Domain via Wikimedia Commons). (CC BY 3.0)

In a similar tone, Thoreau recounts the explosive launching mechanism of witch hazel (*Hamamelis virginiana*) nuts through haptic and sonic remembrance (see Figure 13.3). A witch hazel's woody fruit capsules rupture, releasing two dark glossy seeds as far as 9 metres away. Thoreau brings the 'pretty clusters—clothed, as it were, in close-fitting buck-skin' into his room.[31] 'Three nights afterward I heard at midnight a snapping sound and the fall of some small body on the floor from time to time.'[32] The 'bodies' were the stony seeds of the nuts casting themselves around—an ecological spectacle that continued

Figure 13.3 Flowers and fruits of witch hazel (*Hamamelis virginiana*), Karlsruhe, Germany (2009).

Photo: H. Zell (Public Domain via Wikimedia Commons). (CC BY-SA 3.0)

for three days. This anecdote signifies the convergence of plant materiality (the reproductive strategies embodied in seeds) and affective human memory where fascination, respect and delight predominate. Thoreau's affectedness represents a conception of plant materiality, not as inert substance to manipulate but as animate matter to encounter dialogically: 'I believe almost in the personality of such matter . . . can even worship it as terrene, titanic matter extant in my day . . . we healthily attract one another.'[33] His assertions prefigure Jane Bennett's notion of 'vibrant matter' and the process she describes as 'encountering a vital materiality [in which] *all* forces and flows (materialities) are or can become lively, affective and signaling.'[34] In terms of the environmental conservation value of Thoreau's botanical memory, climate scientists have begun to use his notebooks and memory-based narratives of plants as a 150-year 'dataset' on species abundance and flowering times in researching the effects of climate change on plant species loss.[35]

Although considerably different from Thoreau's nineteenth-century reflections, Seremetakis' chapter 'The Memory of the Senses' also begins with a botanical encounter, specifically with the peaches of her native Greece. For Seremetakis, the content of memory is affected by the physiology of sensory experience—the 'co-mingling' of memory and the senses—mediated by mnemonically resonating things. She asks, 'Is memory stored in specific everyday items that form the historicity of a culture, items that create and sustain our relationship to the historical as a sensory dimension?'[36] Seremetakis understands memory as a 'material practice that is activated by embodied acts and semantically dense objects.'[37] The emotional and historical imbrications borne by such objects 'can provoke and ignite gestures, discourses and acts.'[38] In other words, memory is latent and affective—its dimensions contingent upon, and instigated by, the materiality of objects. Sensory memories of flora—of nibbling, sniffing, scratching—are formed, evoked and deepened through embodied encounters with living plants (such as the springing seeds of witch hazel) and related artifacts (the clothes one wears in a trefoil field or the trekking staff with which one gestures).

Let the Trees Do the Painting: Holly Story's Kino Art

Concerned with the intertwining of memory and flora, I turn now to interviews with southwest Australians. Holly Story was born in 1953 in Zimbabwe and immigrated to Australia in 1971. She is a visual artist and environmentalist who has intensively explored the use of indigenous flora—whole specimens, dyes, fibres and resins—in her artwork. Holly's creative practice employs plant materials to represent and critique settler relationships with the WA environment, particularly its vegetation, which has been historically regarded in acutely polarizing terms: as beautiful and picturesque, as repulsive and worthless.[39] In reference to Holly's exhibition 'Skin Deep' (2006), Stephen Hopper remarks optimistically that 'native plant pigments and their use in dyeing textiles may well signal closer intimacy with plant life from an old landscape.'[40] Holly's art centres around the Deep Creek locality near Walpole, on the south coast of Western Australia, typified by tall open forests of karri (*Eucalyptus diversicolor*) and marri (*Corymbia calophylla*). She calls her work an 'embodied practice'—and an 'attempt to graft myself into the place'—through multisensorial experimentation with the plants in this vicinity.[41]

Holly's Deep Creek studio is nestled in the eucalypt forest, with a sweeping western view across the D'entrecasteaux National Park to the Indian Ocean. In the spirit of Thoreau's cabin, she and her husband have handcrafted their rustic hilltop retreat with a painstaking affection that speaks in the details: alcoves, nooks, a narrow staircase to a sleeping loft. After a short amble in search of the spring's first orchids, Holly shows me her latest work-in-progress, a series of paintings with marri gum. Also called kino, this astringent blood-red exudate is common to *Eucalyptus* trees but most profuse in the endemic marri from which it seeps then congeals to a dark amber-like substance (see Figure 13.4). In the ethnomedicine of the Nyoongar of the southwest, kino has been used externally for cuts, bites and abrasions, and internally for sore throats, diarrhoea and scurvy.[42] According to

Figure 13.4 Close-up of kino (gum) of marri (*Corymbia calophylla*), Kings Park and Botanic Garden, Perth, WA (2011).

Photo: John Charles Ryan.

the nineteenth-century diarist George Fletcher Moore, kino is known as *nalla* in the Nyoongar language.[43] The gum of the endemic tree was an object of curiosity for WA explorers, settlers and visitors. The mariner William Dampier (1651–1715) recorded his gustatory remembrance of kino while navigating the coast of Terra Australis: 'The gum distils out of the knots or cracks that are in the bodies of the trees. We compared it with some gum dragon, or dragon's blood, that was aboard, and it was of the same colour and taste.'[44] In the early twentieth century, novelist D. H. Lawrence (1885–1930), during a visit to Perth, regarded the marri gum in macabre terms: 'leaves and herbage underneath seem bestrewed with blood' and 'this tree seems to sweat blood.'[45]

Holly's engagement with marri unveils the potential of kino, while countering the predominant paradigm of artists appropriating natural materials to produce works. In this example, the trees do the art; the role of the artist is as facilitator, witness and narrative guardian. In her studio, Holly displays large unframed canvases that are encrusted in maroon-coloured gum layers of variable thicknesses. In line with Lawrence's remarks, the topography of kino splotches and splatters of the paintings does invoke blood and guts. The organic patterns provoke a visceral reaction as I imagine the works receding slowly into the soil under the marri trees outside the studio. In contrast, Holly radiates exuberance, playfulness and respect as she conveys her initial encounter with kino as medium. 'I was under one tree putting some paper out just to see what might happen. Then the gum dripped on my head! It was gorgeous. I put my hand up to receive the beautiful red drip. I have tasted the gum. It is sweet when it first comes out of the tree, then it becomes bitter.'[46]

The memory of the sweet-bitter sensory transaction has inspired Holly's continued exploration of eucalypt kino as creative substance. 'I'm now facilitating the creation of artworks by placing things under trees that drip a lot. In fact, some trees drip like taps.'[47] In the co-produced paintings, the linkages between botanical memory (hers, mine and her future audience's), affective modes (from Lawrence's dread to her wonder) and the kino's materiality (the manner in which its physical properties register through sight, touch and taste) coalesce. Holly's work exemplifies the potential for memory to encompass dialogic, embodied encounters between people and plants.[48] Inspired by memory, her artistic experimentations also present an intervention into the historical portrayal of southwest plants, specifically through an enlivened understanding of the material properties of marri kino. The works invite her audience—local and international, conservation-minded and generally interested—to become familiar with or reconsider the potential of *Corymbia calophylla*, as both a medium and an agent in artistic creation. In this instance, the communication of botanical memory through art can transform the popular perception of indigenous southwest vegetation, particularly the eucalypts common to Perth where the majority of Western Australia's human population resides.

A Feeling for the Material: Nalda Searles' Balga Works

Nalda Searles, an artist whose work integrates everyday objects into woven forms, was born in 1945 in the arid Kalgoorlie-Boulder region of WA and presently lives near Perth. Nalda troubles the dichotomies of Indigenous/introduced, cultural/natural and subject/object through her investigation of the Australian environment during a 35-year career. Infused with personal and collective, familial and societal, and human and nonhuman memories, Nalda's art is sympathetic to the materialities of seeds, leaves, fibres, dyes and 'the bush.' She comments, 'My work reflects an awareness of the potential of plants. When I pick up things, because my skills are well-developed, I can almost straightaway identify how they can be used in a creative work.'[49] For Nalda, botanical memory links these materialities to the narratives of plants: 'They have their own indigenous stories. They have colonial stories. They have my story and your story.'[50] Critic Ted Snell has described Nalda's work as 'transformative practice that works with the poetry of materials to reveal what lies beneath the surface, literally and metaphorically.'[51]

Nalda's home studio is replete with previously exhibited works, intriguing found objects, organic miscellany, exhibition catalogues and taxonomic guides, each instigating her memories as I point, ask and listen. Her recollections meander between concepts that inspired works, people who were involved—particularly her mentor Pantjiti Mary McLean[52]—and artistic techniques Nalda has developed or adapted to facilitate the use of WA flora in textiles. Many of Nalda's artworks were conceived during her occasional eight-hour drives from Perth to Kalgoorlie. Motioning towards *Mallee Leaf Jacket* (1996), a salvaged tweed jacket embroidered with eucalyptus leaves, she explains, 'I keep my sharp eye out. I've made that drive at least a hundred times. If there's a storm, a whole branch comes down. If I can reach it, I stop, drag it out of the bush, and put it in the car. Then I have this fresh material. When I get something like that, I need to use it straight away.'[53] Nalda conveys her experience in the present tense, signifying that the memory is not quarantined to the past, but inheres within current moments and future prospects. *The Shape Changers Slippers* (1997), a pair of felted slippers covered in sandalwood (*Santalum spicatum*) sawdust, materializes a memory of her father: 'I had his slippers from when he died. He used bits of rubber called *bowyangs* to hold the slippers on his feet; old men used to do that. Later in his life, he bought a lathe and went bush, like a lot of fellas his age, to collect sandalwood pieces. He'd always offer me the sawdust, which I hoarded.'[54]

Among the panoply of mnemonically laden objects in Nalda's studio, balga bracts—glistening uncannily like an array of dead cockroaches—attract my attention. Known historically as *grasstree* and *blackboy*, *balga* is the Nyoongar name for *Xanthorrhoea preissii*, a tree-like monocot endemic to the southwest (see Figure 13.5). The species name of balga, *preissii*, memorializes Johann August Ludwig Preiss, the nineteenth-century German-born

Figure 13.5 Balga (*Xanthorrhoea preissii*), Yanchep National Park, WA (2007).
Photo: Ausxan, (Public Domain via Wikimedia Commons).

botanist who conducted extensive plant collections early in the history of the Swan River Colony. Bearing tall, thin flowering spikes, balga exhibits grass-like foliage. Its bracts—the specialized leaves of the inflorescence—are dark brown and large. *Blackboy* is an antiquated name that reflects the settler-era likening of the plant to the distant image of an Aboriginal person. The species has been used extensively: for example, dead flower stalks for spear-making, living flower stalks for a fermented drink and resin applied as an adhesive.[55] On setting up camp during an excursion with Nyoongar guides, Moore commented in 1831, 'Blackboy poles are stuck in the ground . . . these are covered with grassy tops of the blackboy: it is a good temporary shelter in rain.'[56] In comparable language, during the 1930s, composer Thomas Wood personified the tree as 'that strange fascination . . . wearing a mop of tousled grass overtopped by a spear.'[57] As the example of balga indicates, colonial and postcolonial resonances inhere in the names of Australian plants.[58] The invocation of a name carries memories of invasion, settlement and the formation of a colonial society through the conversion of ecosystems, including indigenous flora, to pastoral land. Creatively investigating the materiality of southwest plants—in the ways Nalda and Holly do—therefore entails interrogating the relationship between naming practices and the collective memories associated with plants in a place over time. The interconnections between colonialism, memory, naming and plants comprise the subtext of these artworks.

The use of balga bracts in textile art involves personal and collective memories of the plant's cultural significance. For instance, *Balga Blanket* (1995–2008) is a large woollen blanket covered in more than 3000 bracts. Known as appliqué, the technique involved the production of rectangular patterns through the stitching of bracts to the blanket.[59] Another example of appliqué, *Xanthorrhoea Dress* (1996), is a brown woollen gown clothed in bracts and eventually subsumed within the work *Kangaroo Couple* (1995–2008).[60] Made by the artist's mother in 1975, the gown was worn by Nalda for 20 years until its creative appropriation. Similarly sheathed in bracts, *Whiteboy Blazer* (1996) projects the 'colonial connotations' of her work through its satirical play on the obsolete name *blackboy*.[61] Regarding the application of bracts to textiles, Nalda notes her need 'to go inside the plant. I had to put the plant on me to get a feeling for the material. I had already been making big baskets from local flora, but I never stitched plants onto clothing. The idea hit me like a bolt during a drive to Kalgoorlie.'[62] Her tactile use of balga stimulated a distinctly creative period. 'I made *Balga Blanket* during this time. I took a whole double-bed blanket and covered it with balga. I stitched on the bracts.'[63] As mnemonic objects, the bract-based textile works are permeated with the affective registers of balga—a protective plant providing warmth, nourishment, identity and spiritual meaning to Nyoongar people for millennia.

Bitter Tastes, Sweet Smells: Kevin Collins' Banksia Mnemonics

In 1984, Kevin Collins and his family purchased the 'bare grass paddock' in Mount Barker, WA, that would later become Banksia Farm, comprising an arboretum, botanical art gallery and stone guesthouse.[64] Three years on, they had planted nearly half of the known species of banksias (*Banksia* spp.):

> We thought to ourselves, 'we have thirty species and there's only seventy-six [now seventy-eight]. Let's keep going.' We flew to Queensland in 1991 to get the last one, which grows on Hinchinbrook Island. We added new banksias discovered in 2000 and 2007 to complete the collection.[65]

Known for their prominent inflorescences, these shrubs and small trees are native to the Australian landscape; only one (*B. dentata*) occurs naturally outside of the nation and 80 per cent of all species grow only in the south-west.[66] In addition to their role as a food source for native vertebrates and invertebrates, banksias are economically vital to the cut-flower and wild-flower tourism industries.

As conspicuous members of the southwest bush, banksias also pervade the botanical memory of past and present. For instance, bull banksia (*B. grandis*), which is named *poolgarla* in Nyoongar, has been used eth-nobotanically for an array of purposes, including to make a fermented beverage (flowers) and to transport smouldering coals (stalks).[67] During his visit to the King George Sound area, the physician Isaac Scott Nind (1797–1868) confirmed 'every individual of the tribe, when travelling or going to a distance from their encampment, carries a fire-stick. . . . It is generally a cone of *Banksia grandis*, which has the property of keeping ignited for a considerable time.'[68] Moreover, during his 3000 kilometre traverse of the coastline from Adelaide to Albany, the explorer Edward John Eyre (1815–1901) observed, with palpable affect, 'the appearance for the first time of the Banksia, a shrub which I had never before found to the westward of Spencer's Gulf [an inlet in SA], but which I knew to abound in the vicinity of King George's [*sic*] Sound,' the end point of his calamitous expedition.[69] Eyre's 'eagerness and anxiety' were assuaged by the 'degree of satisfaction' he experienced on first spotting banksia.[70] The trees indicated a landscape change, marked his advance to Albany and hence signified the hope of salvation from grim circumstances (mutiny, starvation, extreme thirst). Thus, the colonialist naming of *Banksia* (for the English naturalist Joseph Banks) resonates with memories of explora-tion, mapping, classification and, ultimately, the usurpation of Aboriginal communities from their traditional homelands. In a manner comparable to colonialist naming, the 'ugly little, wicked little' Banksia Men of Aus-tralian author May Gibbs' *Snugglepot and Cuddlepie* (1918), based on the gnarled appearance of old *Banksia* cones, imprints a negative connotation (derived from Gibbs' memory and imagination) on this group of plants, which then attains broader cultural circulation.[71]

In the visitors' centre at Banksia Farm, Collins invokes these diverse sorts of historical narratives as he adroitly selects an object from his table of veg-etal paraphernalia: a *Banksia grandis* cone studded with the woody lips of dehiscent follicles (or dry fruits) (see Figure 13.6). He expounds on its physi-cal properties vis-à-vis Nyoongar understandings of the cone:

> It's the heaviest banksia cone and has excellent qualities. It is very hard. The outer layer smoulders. Aboriginal people transported their campfire with it. In a fresh state, the flowers contain tons of nectar for birds and possums. Aboriginal people made nectar drinks by immersing its flowers in water.[72]

Figure 13.6 Kevin Collins' Table of Curiosities, Banksia Farm, Mount Barker, WA (2009). Photo: John Charles Ryan.

When I ask him about the palatability of its seeds, Collins grimaces then furrows his brow, revealing the ineradicable imprint of a gustatory memory. 'When I first tried a *B. grandis* seed, I spat it out because it was too bitter. However, most banksia seeds are quite pleasant, at least to my palate. In fact, only a couple taste bitter.'[73] Such embodied memories are essential to Collins' instructional approach; he presents his recollections of contact with banksia to foster visitors' memories through the plants' material presence.

During his interactive exposition, Collins tracks seamlessly between sensory memory, banksia physiology and corporeal immediacy. He peers studiously at the follicle of another banksia. 'It is only attached by a small cord to the flower head. You can pop out the follicle with your fingers. Expose it to high heat to get it jumping like popcorn. That will release the seeds.'[74] He encourages me to engross myself sensorially—without inhibition:

Feel the texture. See the old flower heads? If you squeeze them, you will feel the follicles underneath. The structure is very dense, with a strong protective layer, indicating that this species must have fire to release seeds.[75]

Furthermore, human-plant mnemonics involve the interlacing of taste, touch *and* smell. 'This flower smells sweet but you have to put your nose into it. As kids, we would part the flowers, poke our tongues in there, and suck the delicious nectar.'[76] Regarding another banksia's inflorescence, Collins effuses, 'when you squeeze them, little balls of nectar should appear. There's one in the middle there! Just chew the whole flower. You will get some of that nectar!'[77]

Conclusion

During the field interviews reported in this chapter, material prompts (follicles, bracts, kino and derived objects) activated botanical memory to a degree that would not have been possible otherwise. The multisensorial prompts—both living and inanimate, all 'vibrant matter'—discussed in this chapter enabled Ian Smith, Holly Story, Nalda Searles and Kevin Collins to recount (bring to mind) and enact (form new, or deepen) their memories of southwest flora. I assert that these extracts demonstrate that botanical memory is 'sedimented' in the articulations between plants, people and objects. Affectively rich and materially referential, botanical memory also engages empirical understandings of plants, such as the structural characteristics, biogeographical distributions, ecological dynamics and ethnobotanical dimensions of species. These intricate mnemonic narratives are not recited by the interviewees as a series of matter-of-fact, chronologically arranged observations. Instead, their stories are marked by corporeal, spatial and temporal interpenetrations that reach across the past, present and future, while blurring the ontological categories of *human*, *plant* and *environment*.

From these interviews, I believe it is further evident that the proximate senses strongly underlie affective memory. Sucking banksia nectar, touching balga bracts, tasting marri kino and grasping the gimlet staff are corporeal actions with mnemonic resonance. Also apparent in my interviews with artists and collectors are particular attitudes to the sensory remembering of plants through which each individual makes an intervention into the broader communication of botanical memory. For Holly Story, the memories of marri gum that inspire her art also help to dispel the pejorative connotations attributed to kino over history. Similarly for Nalda Searles, the re-imagining of the materiality of balga entails a powerful critique of the colonial residues that underlie the marginalization and exploitation of plants and their Aboriginal heritage. The primary effect of Kevin Collins' leveraging of botanical memory is the broader appreciation of the *Banksia* genus in the experiences of his visitors, who go on to form positive memories of the plants and, by association, the southwest region. In concluding, I suggest that the exploration of plant-based human memory has never been more timely. In an era of catastrophic species loss, the process of sensuous remembering becomes intrinsically linked to the act of conserving.

Notes

1 Ian Smith, interview with John C. Ryan, 29 August 2009, transcript.
2 Ryan 2012, especially Chapter 8; although outside the scope of my discussion, 'botanical memory banking' has been defined by anthropologists as 'the systematic documentation . . . of indigenous practices of local farmers associated with traditional varieties of staple and supplementary crops.' See Nazarea 1998, p. 5.
3 Jones 2007, p. 25.
4 Holly Story, interview with John C. Ryan, 1 May 2014, transcript.
5 Nalda Searles, interview with John C. Ryan, 8 April 2014, transcript.
6 Kevin Collins, interview with John C. Ryan, 9 September 2009, transcript.
7 Piasecka 2014, p. 84.
8 Australian War Memorial 2015.
9 WWF-Australia 2006.
10 Lambers and Hopper 2014, pp. xi–xvi.
11 Ibid., p. xii.
12 Hopper and Gioia 2004, p. 623.
13 Myers 2000, p. 853.
14 Ibid., p. 854.
15 Pate and Beard 1984.
16 Lambers and Hopper, 'Introduction', p. xiv.
17 Hopper and Gioia 2004, p. 644.
18 FloraCultures is an online archive that promotes the heritage of Western Australian plants through art, literature, music, historical writings and oral histories. See, www.floracultures.org.au.
19 See Ryan 2014, pp. 49–58.
20 FloraBase is an open access database of scientific information about all WA plant species. See https://florabase.dpaw.wa.gov.au.
21 Stefano, Davis and Corsane 2012, p. 1.
22 Nazarea 1998.
23 UNESCO 2003.
24 For a discussion of plant ethnography, see Ryan 2012, Chapter 7.
25 Thoreau 1993, 2000.
26 Peck 1990, p. 45
27 Tauber 2001, p. 69.
28 Thoreau 2000, p. 159.
29 Ibid.
30 Ibid.
31 Ibid., p. 190.
32 Ibid.
33 Ibid., p. 168.
34 Bennett 2010, pp. 111, 117 (italics in original).
35 Willis, Ruhdel, Primack, Miller-Rushing and David 2008.
36 Seremetakis 1996, p. 3.
37 Ibid., p. 9.
38 Ibid., p. 7.
39 See, for example, Seddon 2005.
40 Hopper and Gioia 2006, p. 9.
41 Holly Story, interview with John C. Ryan, 1 May 2014, transcript.
42 Clarke 2008, p. 39.
43 Moore 1842, p. 82.
44 Dampier 1703, p. 463.
45 Lawrence and Skinner 1990, p. 93; and Lawrence quoted in Skinner 1972, p. 112.

46 Holly Story, interview with John C. Ryan, 1 May 2014, transcript.
47 Ibid.
48 Holly Story's other work similarly demonstrates the spectrum of modes and practices that could be included within botanical memory. See, for example, *Fancywork*
   (2000).
49 Nalda Searles, interview with John C. Ryan, 8 April 2014, transcript.
50 Ibid.
51 Snell 2009, p. 10.
52 The Aboriginal artist Pantjiti Mary McLean was born in 1930 in Kaltukutjarra,
   Docker River, WA.
53 Nalda Searles, interview with John C. Ryan, 8 April 2014, transcript.
54 Ibid.
55 Ryan 2012, p. 140.
56 Moore 1834, p. 95.
57 Wood 1938, p. 105.
58 The colonial and post-colonial aspects of Australian landscape naming are treated
   extensively by Arthur 2003.
59 To view *Balga Blanket*, see, http://artsearch.nga.gov.au/Detail.cfm?IRN=236056.
60 To view *Xanthorrhoea Dress*, see, http://searlesartist.blogspot.com/.
61 Nalda Searles, interview with John C. Ryan, 8 April 2014, transcript.
62 Ibid.
63 Ibid.
64 Kevin Collins, interview with John C. Ryan, 9 September 2009, transcript.
65 Ibid.
66 Collins, Collins and George 2009, p. 32.
67 Daw, Walley and Keighery 1997, p. 40.
68 Nind 1979, p. 21.
69 Eyre 1845, p. 14.
70 Ibid.
71 May Gibbs quoted in Walsh 2007, p. 76. Indeed, critics of May Gibbs' stories
   commented on the possible effects of this representation of banksia, particularly
   on children: 'One day I received a letter, a very angry letter, saying that it was
   wicked to make drawings of these Banksia Men to frighten the lives out of children' (Gibbs quoted in Walsh 2007, p. 76).
72 Kevin Collins, interview with John C. Ryan, 9 September 2009, transcript.
73 Ibid.
74 Ibid.
75 Ibid.
76 Ibid.
77 Ibid.

References

Arthur, J.M., 2003. *The default country: A lexical cartography of twentieth century Australia*, Sydney: University of New South Wales Press.
Australian War Memorial, 2015. 'Rosemary', available from www.awm.gov.au/commemoration/customs/rosemary/.
Bennett, J., 2010. *Vibrant matter: A political ecology of things*, Durham, NC: Duke University Press.
Clarke, P.A., 2008. *Aboriginal plant collectors: Botanists and Australian Aboriginal people in the nineteenth century*, Sydney, NSW: Rosenberg Publishing.
Collins, K., Collins, K. and George, A., 2009. *Banksias*, Melbourne: Blooming Books.
Dampier, W., 1703. *A new voyage round the world*, vol. 1, London: James Knapton.

Daw, B., Walley, T. and Keighery, G., 1997. *Bush tucker plants of the South-West*, Perth, WA: Department of Environment and Conservation.

Eyre, E.J., 1845. *Journals of expeditions of discovery into Central Australia and overland from Adelaide to King George's sound*, London: T. and W. Boone.

FloraBase, 2016. 'About FloraBase', available from https://florabase.dpaw.wa.gov.au.

FloraCultures, 2016. 'About FloraCultures', available from www.floracultures.org.au.

Hopper, S.D. and Gioia, P., 2004. 'The Southwest Australian floristic region: Evolution and conservation of a global hot spot of biodiversity', *Annual Review of Ecology, Evolution, and Systematics* 35, pp. 623–50. doi:10.1146/annurev.ecolsys.35.112202.130201.

Hopper, S.D. and Gioia, P., 2006. 'Developing through art a botanical and landscape context in a settler society', in H. Story, ed., *Holly story: Skin deep*, Melbourne: SPAN Galleries, pp. 4–9.

Jones, A., 2007. *Memory and material culture*, Cambridge, UK: Cambridge University Press.

Lambers, H. and Hopper, S.D., 2014. 'Introduction', in H. Lambers, ed., *Plant life on the sandplains in Southwest Australia: A global biodiversity hotspot*, Perth, WA: The University of Western Australia Press, pp. xi–xvi.

Lawrence, D.H. and Skinner, M.L., 1990. *The Cambridge edition of the works of D.H. Lawrence: The boy in the bush*, P. Eggert, ed., Cambridge, UK: Cambridge University Press.

Moore, G.F., 1834. *Extracts from the letters and journals of George Fletcher Moore*, M. Doyle, ed., London: Paternoster Row.

Moore, G. F., 1842. *A descriptive vocabulary of the language in common use amongst the Aborigines of Western Australia*, London: W.M.S Orr and Paternoster Row.

Myers, N., et al., 2000. 'Biodiversity hotspots for conservation priorities', *Nature*, 24 February, pp. 853–58.

Nazarea, V., 1998. *Cultural memory and biodiversity*, Tucson, AZ: University of Arizona Press.

Nind, I.S., 1979. 'Description of the natives of King George's Sound (Swan River Colony) and adjoining country', in N. Green, ed., *Nyungar—The people: Aboriginal customs in the Southwest of Australia*, Perth, WA: Creative Research, pp. 15–55.

Pate, J.S. and Beard, J.S., eds., 1984. *Kwongan, plant life of the sandplain: Biology of a South-West Australian shrubland ecosystem*, Perth, WA: The University of Western Australia Press.

Peck, H.D., 1990. *Thoreau's morning work: Memory and perception in a week on the Concord and Merrimack Rivers, the journal, and Walden*, New Haven, CT: Yale University Press.

Piasecka, A., 2014. *Towards creative imagination in Victorian literature*, Newcastle upon Tyne: Cambridge Scholars Publishing.

Ryan, J.C., 2012. *Green sense: The aesthetics of plants, place and language*, Oxford: TrueHeart Press.

Ryan, J. C., 2014. *Being with: Essays in poetics, ecology, and the senses*, Champaign, IL: Common Ground Publishing.

Seddon, G., 2005. *The old country: Australian landscapes, plants and people*, Cambridge, UK: Cambridge University Press.

Seremetakis, C.N., 1996. 'The memory of the senses, Part I: Marks of the transitory', in C.N. Seremetakis, ed., *The senses still: Perception and memory as material culture in modernity*, Chicago, IL: University of Chicago Press, pp. 1–18.

Skinner, M.L., 1972. *The fifth sparrow: An autobiography*, Sydney: Sydney University Press.

Snell, T., 2009. 'Foreword: Nalda Searles: Experience, skill, memory and engagement', in A. Nicholls, ed., *Nalda Searles: Drifting in my own land*, Perth, WA: Nalda Searles and Art on the Move.

Stefano, M., Davis, P. and Corsane, G., 2012. 'Touching the intangible: An introduction', in M. Stefano, P. Davis and G. Corsane, eds., *Safeguarding intangible cultural heritage*, Woodbridge, UK: Boydell & Brewer, pp. 1–5.

Tauber, A., 2001. *Henry David Thoreau and the moral agency of knowing*, Berkeley, CA: University of California Press.

Thoreau, H.D., 1993. *Faith in a seed: The dispersion of seeds and other late natural history writings*, Washington, DC: Island Press.

Thoreau, H.D., 2000. *Wild fruits: Thoreau's rediscovered last manuscript*, B. Dean, ed., New York: W.W. Norton & Company.

UNESCO, 2003. *Convention for the safeguarding of the intangible cultural heritage*, Paris: UNESCO.

Walsh, M., 2007. *May Gibbs, Mother of the Gumnuts*, Sydney: Sydney University Press.

Willis, C., Ruhdel, B., Primack, R., Miller-Rushing, A. and David, C., 2008. 'Phylogenetic patterns of species loss in Thoreau's woods are driven by climate change', *PNAS: Proceedings of the National Academy of Sciences of the United States of America*, 105.44, pp. 17029–33.

Wood, T., 1938. *Cobbers: A personal record of a journey*, London: Oxford University Press.

WWF-Australia, 2006. *The Southwest Australia ecoregion: Jewel of the Australian continent*, Perth, WA: Southwest Australia Ecoregion Initiative.

14 "If I Ever Hear It, It Takes Me Straight Back There"

Music, Autobiographical Memory, Space and Place

Lauren Istvandity

An understanding of our personal selves often comes through reflection on aspects of our identity, experience and our 'life story.' It is through autobiographical memory that we are able to recall and retell of events, moments, people or places that have affected our journey. These recollections are often multi-modal, in that they may consist of one or more sensory elements, any of which may subsequently prompt memories from long ago without warning. These can include visual, olfactory and tactile aspects, or even more abstract discourses of emotion or states of being. Memories can also be *aural*, incorporating a range of sounds and noises, but most strikingly, memories can be filled with, and triggered by, music. This chapter explores the ways in which music can infiltrate autobiographical memories of space and place, and how this can potentially affect the way events are remembered and retold.

This chapter engages the memory narratives of Australians interviewed by the author in a broader qualitative investigation about the interaction between music and autobiographical memory. In one-on-one interviews, 28 adults aged between 18 and 82 were asked to review events in their life as they related to music. Participants were recruited through the use of gatekeepers within the geographical area of southeast Queensland. In all, the cohort comprised 12 males and 16 females, primarily of an Anglo-European or Southeast Asian background. Of the 28 participants, 20 were born in Australia, while eight were born overseas, six of whom spent their childhood in their birth country before emigrating with their family. A semi-structured schedule encouraged participants to describe their involvement with music over their lifetime in generally chronological order. Audible music was not used to trigger memories within interviews, with the exception of one participant who indicated a strong preference to be interviewed with this condition. Of the themes that emerged from interviews in the wider study, connections between music and memories of people (friends and family), emotional times (as occurs, for example, during relationship breakups) and major events (such as weddings and funerals) were most commonly mentioned. Memories of place and space also featured frequently, in varying amounts of detail. Some of the accounts used within this chapter refer to ideas of space and place in oblique ways; however, they are nonetheless considered relevant following Pillemer's assertion that 'memories need not be expressed as explicit, conscious, fully formed narratives in order to be influential.'[1]

Participants regularly offered lengthy and emotionally rich accounts, and while not all interviewees described music as overly important to them personally, the data collectively suggests that memories associated with music, whether music is recorded or experienced live, help individuals to create meaning within their lives via self-reflective practices. Further, it suggests that music helps to capture aspects of space and place within autobiographical memory. The ideas presented in this chapter on connections between musical memories and space or place are supported by narratives that have been filtered from the greater dataset, and therefore are a partial representation of participants.[2] As such, the nature of the following concepts is emphasized as exploratory—a starting point for further investigation.

A Note on Memory

Used across various fields of research, 'autobiographical memory' has come to be typically described in the literature as *memory for events or information concerning the self*.[3] The idea that music can trigger memories is not novel; rather, it is a frequent, everyday occurrence that is commonly ritualized within media consumption; for example, 'golden oldie' radio shows and streaming playlists, 'best of' artist compilations and re-releases, artist comeback tours and so on. It is therefore surprising that, as a topic of academic scrutiny, it has been largely neglected. Of the small body of research that is focused on music and memory, a majority of studies originate in the scientific study of the mind (i.e., psychology, neuroscience, cognitive science, etc.), utilizing quantitative methods to investigate this relationship.[4] At the time of writing, there were scarcely any studies of musically motivated memories that engaged with the topic on a qualitative level, with the exception of studies conducted by Anderson, Hays and Minichiello, and van Dijck.[5]

The greater study from which this chapter draws was based within the field of memory studies, in that it looked to sociological, cultural and psychological theories and research as the basis for an interdisciplinary understanding of the relationship between memory for music and memory for associated events, people and emotions. This investigation contributed further to the sociological understanding of music as a tool for self-reflexive practices that help to make sense of, and allocate meaning to, life experiences. The conceptual framework used in the study privileged the socio-cultural study of memory in everyday life while also incorporating a foundation of scientific memory concepts such as recall,[6] typologies of memory,[7] functions[8] and the ways memory often fails.[9] An appreciation for how the mind works, as provided by such research, can further enrich what we can make of new theories and concepts that consider the implications of autobiographical memory in daily life and society more broadly. In further conceptualizing a framework for the study of musical memory, the author also acknowledges sociological understandings of music, such as those provided by DeNora, Frith, Kotarba and Bennett.[10] In overview, few previous studies from any discipline use in-depth methods to investigate individual experience with musically motivated memory, presenting a clear gap that the broader study hoped to fill. This chapter maintains an interdisciplinary stance, to the end that we may create a more nuanced understanding of autobiographical memory and music through an alternative approach to studying their interaction.

Defining Space and Place

The concepts of space and place as the central tenets of the current discussion require some explanation. The current understanding of these concepts, both separately and in conjunction, appears to hang precariously between, within and across academic fields. Their definitions as explained here, then, are not intended to be widely applicable, but are shaped to this particular context. The ideas of space and place have been approached in this research within a sociological framework; moreover, the description of what is meant by these terms is led by the ways in which research participants detailed their surroundings in memory narratives.

For the purposes of the following analysis, 'place' refers to the physical, tangible elements that combine to constitute an individual's perception of a singular geographical location; it is, as Gieryn[11] describes it, 'a unique spot in the universe.' Ideas surrounding place have attracted a lot of attention in collective memory studies, where physical 'sites of memory,'[12] such as monuments, memorials, archives and the discourse of public memory are examined for their relation to society. In contrast, autobiographical concepts of place are those which are important primarily to the individual, rather than the collective. The concept of 'space' as used in this chapter is less concerned with the tangible,[13] and recognizes more keenly the non-physical aspects of an individual's relationship to the world around them. Following Yi-fu Tuan,[14] space in this context can be thought of as people's spatial feelings and ideas in the stream of experience. In this way, it is not the boundaries of *space itself* that are of interest, but rather the dynamics between self, others and objects that play out within it.

Previous research concerning music, space and place tends to focus on the creation of music within established geographical bounds, such as cities and countries,[15] or on the connections between music, place and identity,[16] although within their work on place, Connell and Gibson do contribute a useful discussion of music consumption and space. In much of this writing, the idea of memory is rather implicit in the broader concepts of identity and background; moreover, these texts predominantly refer to identity and belonging in terms of collective groups of society and culture, and the position and perspective of the individual remains underexposed.

Visual Elements of Memory

A point of significance in establishing the connections between memory, music, space and place is the aforementioned idea that memory often includes sensory elements, such as images, sounds or smells that accompany the information relating to an event or experience. Indeed, Tuan[17] notes the importance of visual perception in his well-known descriptions of space and place as being bound by the senses; it is difficult to think of memories without first recalling visuals. For its ability to most immediately, although perhaps not most effectively, make sense of space and place, recognition of the significant connection between vision and memory, as an addendum to music and memory, is included here.

In his writing on significant event memories, Pillemer[18] refers to the visualization of significant physical places in memory as 'memory landmarks.' This component is crucial to the reconstruction of memory; Pillemer explains that 'constructing a coherent, temporally ordered life history depends on having access not only to the meaning of momentous past events, but also to the imagistic components of personal event memories.'[19] The visualization of place as significant in autobiographical memory, while shared by areas of cultural and collective memory studies, is often taken for granted as part of normal human functioning. Scientific studies of memory have also verified the connection between memory, visual perception and place. Researchers Charis Lengen and Thomas Kistemann note, in the first instance, that the visual cortex is a basic and essential part of spatial perception that helps compose memories, and further, that 'place forms an essential basis for experiences to be unfolded in memory and imagination.'[20]

These researchers place great emphasis on the connection between physicality, place and visual perception within memory: elements of place are therefore often described verbally to 'set the scene' when participants narrate their life experiences. The visualization of space, however, is a much more abstract ideal, and one which was rarely described purposefully by interview participants. While the discussion of space within memory narratives may be somewhat unique to the present context, Pillemer does make note that 'personal event memories cannot be adequately described as occurring within a single level of mental representation or as involving a single mode of expression.'[21] This suggests that memories are not bound by the most immediate elements of physical or visual representation, but may infiltrate greater levels of perception and expression, including atmospheric sensations of space and sound.

The acknowledgement by scholars that sensory information, especially a visual aspect, is usually incorporated into memory suggests that music would form part of the recollected aural environment at the time of the experience. Here then, there is a need to make a note of the breadth of our aural memory, for it is indeed made up of more than just music. Other sounds help us to contextualize our everyday experiences and recollections, just as senses of touch and smell also hold a place in memory. Unlike other atmospheric sounds that might comprise a memory, such as traffic noise, the singing of birds or the chatter in a café, music tends to foster personal associations, which can sometimes be established through mutual interest (e.g., colloquial reference to 'our song' between lovers). Moreover, music, over general sound, can evoke great depths of emotion, and can become illustrative or representational where spoken words do not provide appropriate expression.[22] The inclination for musical sounds to be perceived as personally meaningful to individuals suggests that this particular factor of our aural soundscape may be incorporated into memory in specialized ways.

Memories of Space, Place and Music

This section explores the different ways music, memory, space and place are integrated within select memory narratives from study participants. The analysis queries the association between music and the places or spaces in which it was consumed, and takes into consideration factors that are both intrinsic and external to the memory, including, visual, temporal and contextual aspects. Narratives are presented here through three prominent themes, comprising aspects of personal familiarity, music participation and situational novelty. Though not mutually exclusive, these elements appear to be linked to the significance that space or place may hold in musical memories. These themes are not intended to act as an exhaustive typology, but rather are suggested catalysts for the interaction of music, space and place within the memories of study participants.

Habit and Familiarity

For many participants, the family home frequently formed the backdrop for musical memories from childhood. As a physical place where a considerable amount of time is typically spent, the home often becomes a place of routine, both in terms of the 'everydayness' of activities and the constant feature of the place and space of 'home' throughout our lives, whatever form it may take. Judith Sixsmith describes 'home' as a centre of emotional significance and belonging, as well as being a social unit, a medium of self-expression and identity, and also a base for activity and territoriality.[23] In this way, the home can provide a broader base for meaning-making activities, especially through its typical presence in daily routine. In memory, repetitious experiences within a place or places tend to 'converge,' states Lowenthal, where a 'score of successive scenes soon reduce to one or two.'[24] This is also true where music is involved; indeed, research participants often gave generalized descriptions of childhood memories, with only one or two narratives becoming explicit in their detail.

In the following recollection, the participant describes his association between an oft-watched recording of Queen's *Freddie Mercury Tribute Concert* and the entertainment area of the house:

> George Michael, he got up and did . . . 'Somebody to Love', which is amazing, it's the best version, I reckon it's better than Freddie's version. But they use—like they get a whole gospel choir up there, a black gospel choir up there, it's just amazing. I associate that with the house, the uh house we lived in called Bintaro or in an area called Bintaro [in Indonesia]. . . . So yeah I always associate the floor, the Sony TV that we had . . . Panasonic laser disc player, um yeah, and the red couches that we had which had been with us forever, I sort of associate it with that.
>
> (Matthew, age 25)

Matthew's recollection is a place-based memory: it has a strong visual element that directly accompanies the musical experience in the form of a recorded concert. Leading up to this anecdote, Matthew gave a detailed description of the house itself, the familiarity of the layout of home. The repetition of this activity within this place appears to allow this memory to be recalled vividly through both auditory and visual means.

In the same context, where physical and geographical aspects of place are easily recalled due to their familiarity, this aspect of repetition can also turn memory towards the space and atmosphere of the home environment. Below, Will's narrative is focused on the dynamic of a particular room in his childhood home:

> I always remember, like Sundays for me are synonymous with Beethoven or Tchaikovsky or Bach, maybe some Grieg . . . just on a Sunday, my Dad reading the paper. That's from an early age, every single Sunday there was classical music on, early morning. And sit and read the paper so, definitely in my new house, yeah since I was ten. Don't necessarily remember that beforehand, but from ten, the nice room of the house, that sort of Sunday, you go in there, especially in summer. So . . . now because of it, like today—no one was in the house and I feel that it's a Sunday, even though it's a public holiday Monday, I will put on classical music, because I don't know, it's a nostalgia thing or whatever.
>
> (Will, age 26)

Here, Will relates very much to the atmosphere or space within the house. The music triggers a sense of time, extending to the season, the day of the week, as well as the time of day; temporality therefore plays a large part in forging the circumstances of the event. While tangible, visual elements are still present in the memory (the 'nice' room, the newspaper), Will's attention is more focused on the way in which a mood was created, but only as a result of particular additives. The fact that this music (and its required technology) is accessible for Will means that he can authentically reproduce the space within this memory, especially at times when his current physical environment resembles that of the original experience.

Familiar places and spaces, and habits within them, appear to foster an interaction between memory and music. It is also interesting to note that the participants in the above narratives also appear to be well acquainted with the music that was present, suggesting that such memories are formed through repeated experiences with both music *and* space or place. Lowenthal's idea of convergence can again be highlighted: although it is likely that minor variances occurred in these elements upon each encounter (e.g., Will notes different composers feature in his memory), the repetition of both music and surroundings results in one comprehensive narrative.

Embodiment and Participation

Besides the act of listening to music, activities that produce or accompany music may also influence the ways musical memories are created and recalled. That is to say, singing or playing music, dancing or otherwise moving to music are other aspects that can tie music to memories of space and place. As I detail elsewhere,[25] the embodiment of music through dance or movement appears to further enhance the ability of music to encapsulate memory details.

This was exemplified by a participant who completed our interview with the aid of a playlist of his favourite music, which he had compiled for his recent sixtieth birthday celebrations. Upon hearing 'The Twist,' Ian recalled this memory:

> This was when I would have been about eight or nine I guess. [It] was revolutionary when 'The Twist' came out because it was so different . . . and it had a dance that went [with] it. . . . And I can remember it because it was, we worked on the farm for these people . . . they were rich land owners you know, and they . . . had this brand new stereo, we'd never seen a stereo before, and it was on this huge cabinet, it was about this big here and it was dark mahogany. . . . When they weren't there we would dance in their lounge to The Twist . . . that was quite significant.
>
> (Ian, age 60)

Within our interview, Ian had referred to the financial disparity between the farm owners for which his family worked and his parents' own position. Hence, this narrative conveys some feeling of irreverence at dancing in the house without the owners' knowledge. That the family was also dancing to 'The Twist,' with its subtle undertones of hedonistic or rebellious behaviour only serves to further intensify the significance of that lounge room space in

Ian's memory. Like Matthew's recollection of the lounge room in his Indonesian house, the description of tangible items, such as furniture, help to make this a place-based musical memory.

For some participants, the idea of 'home' (as mentioned earlier) differed from the greater norm of suburban family living. In Vincent's case, much of his life prior to adulthood was spent at boarding school. In Vincent's stories, family members featured rarely, and memories contextualized by college grounds were stronger than memories of 'home.' Nonetheless, Vincent had a number of musical experiences within senior college that have endured as some of his most vivid memories, such as this one in the school chapel:

> I survived Nudgee Junior and then in 1949 I . . . became a scholarship student at Nudgee Senior at Nudgee College, Boondall. Life was different there; they were more civilized. They did have a magnificent chapel and they did have a practice of singing every Sunday, a very, shall we say, broad range of hymns, 'Faith of our Fathers,' 'Soul of my Saviour.' I can never hear any of those without being back in that chapel, because that's the first time I ever had any participatory music where you're doing it with a lot of folk you know. Later on they formed the formal choir, which they had to lead in high mass, which was celebrated about three times a year. And that's when we sang Joseph Smith's Mass in C, which had part-singing and all sorts of things, I don't hear it at all any more but if ever I do hear [sings in Latin]—immediately I'm back in that chapel.
>
> (Vincent, age 77)

This narrative presents a strong connection between place and musical participation. Vincent described emphatically the resulting visualization that can be triggered upon hearing particular hymns, which is tied to the sensations of group singing. The physicality of singing can be thought of as producing a different kind of meaning within musical memory: Vincent himself was invested in producing the sounds with which he associated the college chapel, and in this way was more intrinsically connected with this memory.

Various progressions in technology throughout the twentieth and twenty-first centuries had affected the ways in which participants consumed and interacted with music, which in turn can affect the way music is memorialized by individuals. The effects of music technology were often seen as having the greatest impact upon participants' musical experiences within cars. While the physical properties of a car could lead us to think of it as a 'moveable place,' it can be more easily understood in terms of space, where the dynamic of the zone is changed by the interaction of elements within it. Contributing to research specifically on music interaction within the car, Michael Bull also defines the car as a 'space' when he states: 'The car is a space of performance and communication where drivers report being in dialogue with the radio or singing in their own auditized/privatized space. . . . It is simultaneously private and public.'[26] The act of listening to, or where music technology is lacking, producing, music within the car can therefore be thought of as further interweaving space, music and memory.

In the following narrative, Bea fondly remembers car trips from rural Queensland to the capital city Brisbane, in which recorded music or radio was absent:

> My Mum's name was Kathleen, and we used to sing all the way to Brisbane of course, in the car, but then Dad always sang, ah, for mum [pause, Bea is getting a little upset]—that's silly. Ah yes, he always sang 'I'll Take you Home Again Kathleen' and Dad always sang that, that was the only song he ever sang, and we all kept quiet in the car while Dad used to sing it. Oh, they had a wonderful married life.
>
> (Bea, age 81)

Here Bea's father commands the social and musical space of the vehicle—his contribution to singing, while singular, was attributed more respect than other family members' songs through the family's silence. Bea's emotional recollection is framed through the act of participation, however in this instance it is the absence of her own singing that is significant. This memory also contains aspects of habit, and is an example of Lowenthal's 'convergence,' where successive versions of this experience amount to one over-arching narrative.[27] The idea of the car as a 'space' is evident here through an emphasis on the dynamics of control between family members. The music, especially that particular song, is ingrained in memory, not only due to the presence of family within the car but also due to the replication of family members' status within it.

Novelty, Vividness and Impact

In some instances, the presence of unusual or novel aspects within experience can result in particularly strong or vivid memories.[28] In research participants' narratives, direct relations to place or space sometimes were the focal point of such memories. More often, space or place provided the platform on which vivid memories were played out. Although vivid memories are well documented, their recollection in relation to music has received limited attention in academic accounts. Vivid moments can sometimes also be instructive for the individual, which Pillemer (2001) refers to as a 'memory directive.'[29] One such example was given by Paul, who described the first time he heard 'The Saturday Boy' by Billy Bragg:

> ['The Saturday Boy'] is basically an unrequited love song and in the middle of it there's a gorgeous trumpet or flugelhorn solo, so there's this beautiful contrast between this real roughness of the guitar playing, the

kind of almost clumsy song lyrics and then this quite sublime trumpet solo in the middle of the piece. . . . And um, I must have been 15 or 16 and after a night of drinking at a friend's place, drinking cask wine and that kind of thing. He had an older brother who came home and who had just been to a sex worker. And so I was having a good time kind of drinking with my friend . . . and then his brother came home and he was a bit of a cave man and he was describing to us what he'd done with this sex worker . . . and I remember feeling deeply uncomfortable in this situation, thinking 'oh I'm a long way from home and I really just want to go home. It doesn't seem quite right, this seems quite rough to me, like I don't really belong here' and anyway um then, that song was put on and I don't think it was communally put on . . . it just happened to be on in the background and that trumpet solo came and I thought, well—I was very moved by the music at that time. It seemed quite sublime even in its roughness, so there's something about the very particular situation, I could draw the room we were in actually if necessary like a map of it anyway, and, yeah . . . from that moment I felt well, beauty can come from the most awful ugly places, that it's possible to connect with your friends very deeply over music.

(Paul, age 42)

Paul's narrative incorporated both visual and aural aspects that contributed to the intensity of the memory. The jarring contrast between the awkwardness of the situation and Paul's perception of the music as beautiful is mirrored in his description of the song itself. Paul's physical surroundings served as a platform for the scenario, drawing together place and music. The coincidental nature of the situation—the graphic story and the song playing in the background—helped to produce greater meaning for Paul, who can recall great detail in association with this song, including the physical layout of the house.

Although autobiographical memory is memory for the self, it frequently crosses paths with collective memory. The concept of collective memory, broadly defined, refers to the 'representation of the past, both the past shared by a group and the past that is collectively commemorated, that enacts and gives substance to the group's identity, its present conditions and its vision of the future.'[30] In the following extract, Vincent shares his experience of an event that affected, and was subsequently memorialized, by people around the world:

The first music I ever remember was on V.E. Day in 1945, in August,[31] when the Japanese surrendered and I was in Nambour [Queensland]. It was holidays, I walked up the street and down the street—now you have to remember . . . from the mid-30s 'til the mid-40s the only source of any music generally speaking was the ABC,[32] on a wireless if you happened to have a wireless, and not everybody had a wireless . . . so we didn't have a lot of music. But walking up the street, every shopfront doorway I went to, there were the young women who worked in there, and they were all singing the same song, and this really impressed me, and the same song was in my mind when I started talking and now it's faded out for a second. They'd linked arms and they were waltzing back and forth, waltzing back and forth kicking their legs up and doing everything else . . . but every time I heard that song I was immediately back on V. E. Day, walking out Currie Street, Nambour.

(Vincent, age 77)

Later in the interview, the song came back to Vincent—it was 'The Hokey Pokey.'[33] This narrative illustrates the personally affective detail with which an individual can memorialize an event that is collectively experienced. This particular instance is further characterized by this particular song, which for Vincent heralded a memory composed of visual, aural, geographical and emotional components.

Making this memory especially acute for Vincent was the discourse of novelty, which is present in a number of ways. Recorded music was not frequently heard during the war years—the unfamiliar presence of music, and overtly happy music at that—is one of the first things that caught Vincent's attention in situ. The subsequent understanding of the significance of the music—that the war was at an end—can also be thought of as novel, in that it was a long hoped-for, but essentially rare event. The memory is also firmly located within place—Vincent remembers the town, the street and the shopfronts that physically bind the experience. The tie between place and music in this memory is also further ingrained through the embodiment of music and emotion: people in the memory are dancing and singing, the visual of which he could easily recall.

Conclusion

In this chapter, music, space and place have been shown to interact in a range of ways within autobiographical memory narratives. As part of a larger project, one of the first to look at autobiographical memory from a qualitative, socio-cultural angle, this investigation contributes to a more nuanced understanding of how music and memory interact within circumstances of space and place in daily life. Though it is exploratory in intention, this inquiry reveals that some catalytic elements of experience tend to increase the significance, and therefore, the inclination to recall space- or place-based memories that involve music. Where individuals repeatedly encountered space, place or music that was familiar or habitual to them throughout life, recollections tended to manifest as an over-arching, converged memory, rather than as single events upon re-telling. The recurrence of scenarios could also result either in the recognition of physical and geographical elements within narratives, or contrastingly, participants also recalled associations with spaces, where tangible aspects were overridden by particular atmospheric sensations. On the other hand, elements of bodily involvement with music can further intertwine space or place with musical experience. Reasons for human proclivity to move to or produce music remain under debate; however, in this context it seems that the personal investment in the music within particular space or place boundaries can further cement the tie between these elements in autobiographical memory. Finally, memories containing unusual or novel factors also underlie significant music, space or place recollections. Incorporating vivid visual and aural aspects, these memories also tend to strike the individuals as instructive or important to their life story, where music becomes essential in its re-telling. These themes are in no way exhaustive, but are suggestive of the types of contexts in which music, space and place interact within memory, presenting a starting point for further exploration into the social, cultural and geographical components of autobiographical memory.

Notes

1 Pillemer 1998, p. 23.
2 Pseudonyms are used for participants throughout this chapter.
3 Brewer 1988; Conway and Plydell-Pearce 2000; Fivush 2011.
4 For example, Schulkind, Hennis and Rubin 1999; Janata, Tomic and Rakowski 2007; Cady, Harris and Knappenberger 2008.
5 Anderson 2004; Hays and Minichiello 2005; van Dijck 2006, 2009.
6 For example Rubin, Rahal and Poon 1998.
7 For example Brown and Kulik 1977; Pillemer 1998, 2001.
8 Cohen 1996; Alea and Bluck 2003; Fivush 2011.
9 For example Schacter 2001; Loftus 2005.
10 Frith 1981, 1987; Bennett 2000, 2013; DeNora 2000; Kotarba 2002.
11 Gieryn 2000, p. 464.
12 Nora 1989.
13 Gieryn 2000.
14 Yi-fu Tuan 1979.
15 For example Cohen 1995; Forman 2000.
16 For example Connell and Gibson 2003; Whiteley, Bennett and Hawkins 2004.
17 Tuan 1979.
18 Pillemer 2001, p. 96.
19 ibid.
20 Lengen and Kistemann 2012, p. 1169.
21 Pillemer 2001, p. 22.
22 See Horton 1957.
23 Sixsmith 1986, pp. 281–82.
24 Lowenthal 1975, p. 28.
25 Istvandity 2014.
26 Bull 2004, p. 249.
27 Lowenthal 1975.
28 Brown and Kulik 1977.
29 Pillemer 2001.
30 Misztal 2003, p. 25.
31 The participant referred to 'Victory in Europe Day,' which marks officially the end of World War II, celebrated in Australia on 8 May 1945. Given his description, it is

more likely that he is referring to V.J. Day (Victory over Japan Day, also called V.P. Day, for Victory in the Pacific Day), which occurred on 14 August 1945.
32 Refers to the national public radio and news broadcaster, the Australian Broadcasting Corporation.
33 A traditional song with gestures implied in the lyrics.

References

Alea, N. and Bluck, S., 2003. 'Why are you telling me that? A conceptual model of the social function of autobiographical memory', *Memory* 11(2), pp. 165–78. doi:10.1080/741938207.

Anderson, B., 2004. 'Recorded music and practices of remembering', *Social and Cultural Geography* 5(1), pp. 3–20. doi:10.1080/1464936042000181281.

Bennett, A., 2000. *Popular music and youth culture: Music, identity and place*, Hampshire, NY: Palgrave.

Bennett, A., 2013. *Music, style, and aging: Growing old disgracefully?* Philadelphia, PA: Temple University Press.

Brewer, W.F., 1988. 'Memory for randomly sampled autobiographical events,' in U. Neisser and E. Winograd, eds, *Remembering reconsidered: Ecological and traditional approaches to the study of memory*, Cambridge, New York and Melbourne: Cambridge University Press, pp. 21–90.

Brown, R. and Kulik, J., 1977. 'Flashbulb memories', *Cognition* 5, pp. 73–99. doi:10.1016/0010–0277(77)90018-X.

Bull, M., 2004. 'Automobility and the power of sound', *Theory Culture Society* 21, pp. 243–59. doi:10.1177/0263276404046069.

Cady, E.T., Harris, R.J. and Knappenberger, J.B., 2008. 'Using music to cue autobiographical memories of different lifetime periods', *Psychology of Music* 36(2), pp. 157–78. doi:10.1177/0305735607085010.

Cohen, G., 1996. *Memory in the real world*, East Sussex: Psychology Press.

Cohen, S., 1995. 'Sounding out the city: Music and the sensuous production of place', *Transactions of the Institute of British Geographers* 20(4), pp. 434–46.

Connell, J. and Gibson, C., 2003. 'Sound tracks: Popular music, identity and place', London: Routledge.

Conway, M.A. and Pleydell-Pearce, C.W., 2000. 'The construction of autobiographical memories in the self-memory system', *Psychological Review* 107(2), pp. 261–88. doi:10.1037/0033–295X.107.2.261.

DeNora, T., 2000. *Music in everyday life*, Cambridge: Cambridge University Press.

Fivush, R., 2011. 'The development of autobiographical memory', *Annual Review of Psychology* 62, pp. 559–82. doi:10.1146/annurev.psych.121208.131702.

Forman, M., 2000. ' "Represent": Race, space and place in rap music', *Popular Music* 19(1), pp. 65–90. doi:10.1017/s0261143000000015.

Frith, S., 1981. *Sound effects: Youth, leisure, and the politics of rock'n'roll*, New York: Pantheon Books.

Frith, S., 1987. 'Towards an aesthetic of popular music', in R. Leppert and S. McClary, eds, *Music and society: The politics of composition, performance and reception*, Cambridge, New York and Melbourne: Cambridge University Press, pp. 133–49.

Gieryn, T.F., 2000. 'A space for place in sociology', *Annual Review of Sociology* 26, pp. 463–96. Available from: http://www.jstor.org/stable/223453.

Hays, T. and Minichiello, V., 2005. 'The meaning of music in the lives of older people: A qualitative study', *Psychology of Music* 33(4), pp. 437–51. doi:10.1177/0305735605056160.

Horton, D., 1957. 'The dialogue of courtship in popular songs', *American Journal of Sociology* 62(6), pp. 569–78. Available from: http://www.jstor.org/stable/2773132.

Istvandity, L., 2014. 'The lifetime soundtrack: Music as an archive for autobiographical memory', *Popular Music History* 9(2), pp. 136–55. doi:10.1558/pomh.v9i2.26642.

Janata, P., Tomic, S.T. and Rakowski, S.K., 2007. 'Characterisation of music-evoked autobiographical memories', *Memory* 15(8), pp. 845–60. doi:10.1080/09658210701734593.

Kotarba, J., 2002. 'Rock 'n' roll music as a timepiece', *Symbolic Interaction* 25(3), pp. 397–404. doi:10.1525/si.2002.25.3.397.

Lengen, C. and Kistemann, T., 2012. 'Sense of place and place identity: Review of neuroscientific evidence', *Health & Place* 18, pp. 1162–71. doi:10.1016/j.healthplace.2012.01.012.

Loftus, E.F., 2005. 'Planting misinformation in the human mind: A 30-year investigation of the malleability of memory', *Learning and Memory* 12, pp. 361–66. doi:10.1101/lm.94705.

Lowenthal, D., 1975. 'Past time, present place: Landscape and memory', *Geographical Review* 65(1), pp. 1–36. doi:10.2307/213831.

Misztal, B.A., 2003. 'Theories of social remembering', in L. Ray, ed, *Theorizing society*, Maidenhead, UK and Philadelphia: Open University Press.

Nora, P., 1989. 'Between memory and history: Les lieux de mémoire', *Representations* 26, pp. 7–24. doi:10.2307/2928520.

Pillemer, D.B., 1998. *Momentous events, vivid memories*, Cambridge, MA and London: Harvard University Press.

Pillemer, D.B., 2001. 'Momentous events and the life story', *Review of General Psychology* 5(2), pp. 123–34. doi:10.1037/1089-2680.5.2.123.

Rubin, D.C., Rahhal, T.A. and Poon, L.W., 1998. 'Things learned in early adulthood are remembered best', *Memory & Cognition* 26(1), pp. 3–19. doi:10.3758/bf03211366.

Schacter, D.L., 2001. *How the mind forgets and remembers: The seven sins of memory*, Boston and New York: Houghton Mifflin Company.

Schulkind, M.D., Hennis, L.K. and Rubin, D.C., 1999. 'Music, emotion and autobiographical memory: They're playing your song', *Memory & Cognition* 27(6), pp. 948–55. doi:10.3758/bf03201225.

Sixsmith, J., 1986. 'The meaning of home: An exploratory study of environmental experience', *Journal of Environmental Psychology* 6, pp. 281–98. doi:10.1016/s0272-4944(86)80002-0.

Tuan, Y.-F., 1979. *Space and place: Humanistic perspective*, Netherlands: Springer.

van Dijck, J., 2006. 'Record and hold: Popular music between personal and collective memory', *Critical Studies in Media and Communication* 23(5), pp. 357–74. doi:10.1080/07393180601046121.

van Dijck, J., 2009. 'Remembering songs through telling stories: Pop music as a resource for memory', in K. Bijsterveld and J. van Dijck, eds, *Sound souvenirs: Audio technologies, memory and cultural practices*, Amsterdam: Amsterdam University Press, pp. 107–19.

Whiteley, S., Bennett, A. and Hawkins, S., 2004. *Music, space and place: Popular music and cultural identity*, Surry: Ashgate.

15 Seeing in Black and White
Visualizing "Shadow Sisters" among Metaphors of Light and Dark

Emma Dortins

In Judith Wright's papers, folded in with sympathy cards and letters she received in the months following the death of her friend Kath Walker, is a small circular piece of white card marked in black felt-tip pen, forming two female faces in profile.[1] The card is captioned simply 'shadow-sisters.' This was an expression that the two friends used to characterize their friendship over 20 years as they acted out testimony and listening, recognition and forgiveness, sharing and difference, in a partly public narrative that explored possibilities for cross-cultural social healing. This visual embodiment of the expression captures aspects of the friendship as represented by the two women themselves. At the same time, it brings into view the ways in which the friends drew on the metaphor of black and white and engaged with and promoted this visuality of race.

Wright grew up on a family property on the New England tablelands of northern NSW, where her family had farmed since the 1840s, the family pioneer having migrated from Wiltshire in 1827.[2] As a child, she was isolated, with two younger brothers engaged in farm work and her closest friend, a cousin, 30 kilometres away. Her mother, in ill health until her death when Wright was 11, encouraged her to write poetry. Wright's father heeded her pleas to attend university, and she moved to Sydney to study literature, philosophy and psychology in 1934, not yet quite 20. Returning to the family property to help her father during the war, she began to write with vigour, and was first published in 1942.[3] As she wrote, she began to reflect on her family's involvement in the dispossession of the Wadja people, and the clearing and exploitation of their land, two intertwined histories which were to remain lifelong preoccupations.[4]

Born in 1920, Walker grew up on her father's Noonuccal country on North Stradbroke Island, off southern Queensland, as part of a community whose descendants have recently achieved a successful Native Title determination over parts of the island and its surrounding waters.[5] Following a typical path for young Aboriginal women of her generation, she left school and went into domestic service in nearby Brisbane at the age of 13, hardworking and underpaid, with minimal education. When other opportunities presented themselves, she took them. She gained further education via army service, and after the war joined the Communist Party and its Realist Writers movement.[6] By the late 1950s, she was an experienced speaker on local issues, and began to involve herself in wider campaigns.

When the two women crossed paths they were both in their forties. Wright had five major publications under her belt, and was already recognized as one of Australia's greatest poets.[7] She was involved with the nascent conservation movement as a founding member of the Wildlife Preservation Society.[8] Walker was deeply involved in the civil rights movement and was one of the Federal Council for the Advancement of Aborigines and Torres Strait Islanders' most valuable campaigners. She spoke and lobbied tirelessly for a 'yes' vote in the successful 1967 referendum, which allowed the national government to make laws for Aboriginal and Torres Strait Islander people and ended their exclusion from the census.[9] When *We Are Going* was published in 1964, she also became nationally recognized as a poet. A friendship blossomed which Wright later felt had changed her whole orientation towards the world, helping her to break through a deep-seated, subconscious reaction to Aboriginal people, 'a sort of shame mixed with misunderstanding' that she could now see in all settler Australians.[10]

After living and working in Brisbane for many years, Walker returned to live on North Stradbroke Island in 1971 and established a cultural and education centre known as Moongalba. In 1987, she officially changed her name to Oodgeroo of the Noonuccal Tribe.[11] Wright, after a long period of intense mourning for her husband Jack McKinney, moved from their home near Brisbane to Braidwood in 1975, over 1000 kilometres to the south.[12] The friends' private relationship was conducted mostly via letter—especially as Wright's hearing deteriorated. Their public friendship was imaged in their poetry, and via their public speaking circuits, interviews for magazines and occasional radio and television appearances, leaving an important legacy to champions of reconciliation.[13]

—

The shadow sister card is signed simply 'JS.' The artist was possibly Judith Sercombe, whose letter to Wright is enclosed in the same folder. Like a number of other correspondents writing to offer their sympathies, Sercombe did not know Walker or Wright personally, but had been touched by the friendship (she had seen Walker talking about it on television some years earlier) and felt compelled to make contact with Wright on the loss of her friend. Sercombe was a ceramicist, and had been inspired by Walker's television appearance to create a 'two faced' ceramic doll with a 'shadow sister' theme, a similar artistic project to that of the card.[14] Whoever put pen to paper was one of many who wrote to Wright wanting to pay tribute to the profundity, longevity and equality that they saw in this friendship.[15] Sentiment was high, and the cross-cultural or cross-racial nature of the friendship was its key feature in the minds of several correspondents. A young woman who had helped Walker put her papers in order over a number of months wanted Wright to know: 'we talked about the divisions of race being leaped by love and the luminescent quality that this love has brought to our lives . . . I know that you sustained and inspired and loved one of the most brilliant people to ever live.'[16]

The shadow sister card brings together two visual metaphors, both based in the contrast between light and the absence of light. 'Shadow sister' was a term coined by Wright and used by the friends themselves as a term of solidarity and endearment as well as being adopted by others in representing their friendship. The card crystallizes this epithet into visual form. By using the white of the card in contrast with the black of the pen, the artist has created two female profiles, facing inward towards each other, one inverted while the other is upright and vice versa, lips slightly parted as if in conversation. In doing so, as well as invoking something of the dynamism and equality of the friendship, the card draws the two friends into a second metaphor of light and dark: the ubiquitous, the powerful and often reductive signification of racial difference through 'black and white.' Over several years of thinking about this friendship, and its representation, I had wanted to turn the sensitive, home-made 'shadow sister' away from this problematic leviathan. I had told myself that it was simply this artist's interpretation, and wondered how many others had visualized 'shadow sister' in this way. But looking back at the language used by Walker and Wright themselves, I realize that black and white was a central working concept for both women, a concept that had brought them together and sustained their conversations over the years. The black-white visuality of race was not something they sought to avoid; instead they succeeded in investing the metaphor with a positive meaning that would perhaps help to form a basis for the imagery of reconciliation.

The term 'shadow sister' became a well-known epithet for the friendship. It was perhaps first used by Wright in conversation with Walker when she stayed with her in the fashion of a writing retreat in the summer of 1971/72. Almost mystically, the two women 'recognized' the short story Walker was then working on, titled 'Oodgeroo,' in Wright's poem 'Canefields Country,' first published in 1955, long before the two women had met. Grown to womanhood, the young girl of Wright's poem begins to inscribe the waiting paperbark parchments with the knowledge of the land in Walker's story. As Walker later recalled the encounter, when she had asked Wright how she wrote that poem, 'how did Judith *know*?' Wright had replied, 'I think you and I have met a long time ago in yet another dreamtime when we were shadow sisters.'[17] In this encounter, as Walker understood it, the two women were able to meet on a number of levels: personally, creatively and on a level which defied time and place as well as being thoroughly situated in them.

Following this visit, Wright wrote a poem for Walker, titled 'Two Dreamtimes,' in which she addressed Walker as her 'shadow sister.' In the terms of the poem, this meant that the two women came together from polar opposites of Australian historical experience—'I am born of the conquerors, / you of the persecuted'—to share a sisterhood of activism, hope and loss. Where Wright's forbears stole the country of Walker's Aboriginal ancestors, now the women shared a loss of country—to extractive industry, foreign investment and the government's disregard for a future beyond short-term profit—and of enchantment: 'If we are sisters, it's in this— / Our grief for a lost country.'[18]

The poem was first published in Wright's 1973 collection *Alive*, and has been widely commented on as a characterization of the friendship from Wright's point of view.[19] The poem enjoyed a long life as part of their public friendship—Walker read it, for example, as part of her acceptance speech, Honorary Doctorate of Letters, at Griffith University in 1989, and Wright read it at a vigil outside Parliament House to mark the thirtieth anniversary of the 1967 referendum.[20] 'Shadow sister' became a term widely known to be associated with the friendship, forming, for example, the title of Frank Heimans' 1976 film biography of Walker (in which Wright briefly appeared).[21] Already in the poem, 'shadow sister' was a form of togetherness that overthrew the effective apartheid of Wright's youth, 'You were one of the dark children / I wasn't allowed to play with— / riverbank campers, the wrong colour, / (I couldn't turn you white).'

Wright's interest in the 'shadow' was informed by Jungian principles, in which the shadow is a part of the unconscious, cast onto the world as we try to perceive it, and encountered by each person as a part of that external world, rather than as a part of herself.[22] In her relationship with Walker, and in her writing through the 1970s in particular, Wright was undertaking that 'considerable moral effort' Jung felt was necessary to make the shadow conscious and remove this barrier to emotional maturity. Specifically, Wright strove to recognize her culturally inherited fear of Aboriginal survival and to put it aside, so she could acknowledge Aboriginal sovereignty.[23] As Jennifer Jones observes, Wright 'eschewed the conventional and convenient invisibility of whiteness, choosing instead to face her inheritance.'[24] In Wright's work, the Jungian concept of the shadow immediately became entangled with the visuality of race—Wright wanted to focus clearly, dispelling the mists of the fearful, self-protecting shadow in favour of a bright light on the realities and relationships of black and white.

Applied to European Australians' fears about the survival of Aboriginal people in the face of oppression, the Jungian 'shadow' absorbed some of the moral content of 'black and white' that had developed over centuries and been used to great effect as racial ideas hardened in the nineteenth century: a juxtaposition of primitive 'darkness' with a literally en-lightened modern Western culture.[25] Wright's poem 'The Dark Ones' (1976) harnesses the imagery of night and day to the shadow, with an Aboriginal presence looming or haunting by night the town which is a safe, successful, homogeneously white settlement by day. One of Wright's foremost appreciators and critics, writing in the late 1970s, enthused about the connections between shadow and primitive she found in this poem where Aboriginal people were 'identified with the dark and potent contents of the unconscious . . . the shadow side of the self,' rising 'up like wraiths to confound and reproach the confidence and assurance of the daylight world.'[26]

'Shadow *sister*,' however, broke free from these implications. 'Sister' denotes an imagined kinship of solidarity, women working united, borrowed from second-wave feminism and religious women's activism, a comradeship which was very much a part of the friendship. In 'Two Dreamtimes' Walker is no wraith, but is real and irreducible, with a 'knifeblade flash' in her eyes. She strides briskly out from the shadows of regret and guilt, to share grief, a love of the land and the life of a poet with Wright, and to teach her something of the reality of Aboriginal experience. Walker adopted the term. She made it mutual, addressing Wright as *her* shadow sister in a poem-in-reply, 'Sister Poet' (1975). Here, she reproduced the poles that Wright had set up, writing back from her 'sit down place with the Koorie dead,' but joined them in a single sphere of understanding, as the two women bound together by the 'shadow' share their thoughts and dreams without reservation.[27]

The way in which Walker had responded and adopted the term lent itself to equivalency, as if the two women were seeing reflections of themselves and their people's histories in each other. The two profiles on the 'shadow sister' card—one upright when the other is inverted and vice versa—are not unlike the two complementary energies of the Chinese yinyang symbol, which balance each other in a dynamic perpetuity. The image suggests parity, closeness, difference on a common ground.

—

'Black and white' is a verbal appeal to the visual sense—one of the basic contrasts that governs our visual perception of the world—light and the absence of light. 'Visuality,' in the sense that it is developed by Whitney Davis, refers to a way of looking culturally, so that the seen is much more than the literally seen. The content of the image is held in the viewer, enculturated to see more than he or she is looking at.[28] The 'shadow sister' card engages its viewer in such a cultural act of seeing, in recognizing Walker and Wright and their 'cross racial' relationship in its stylized black and white image.

As well as being part of an ancient global language of race, black and white has a strong, ongoing local presence as a way of designating contiguous difference between Aboriginal and Torres Strait Islander people and settler Australians. Yin Paradies writes of the 'Black–White racial dichotomy so fervently clung to in Australia' and its oppressive implications for Indigenous people whose appearances don't correspond with the expected 'Indigenous look.'[29] Popularly, the problems of the metaphor are highlighted amidst its ongoing reproduction. A 2005 project co-organized by 'Dare to Lead,' an Indigenous education project, and Reconciliation

Australia asked selected Indigenous and non-Indigenous students from six secondary schools to produce photographic images communicating about reconciliation. Many of the images juxtaposed black and white props (shoes, umbrellas), as well as contrasting-coloured skins, to celebrate reconciliation. But the project also generated a strong critique. A photograph of two young women submitted by a student in Alice Springs has been interpreted by teachers and students as 'black and white coming together as equals' when actually it is a portrait of two Aboriginal cousins taken by a third cousin with the aim of challenging this very assumption. A young Aboriginal man from Maroochydore was angered to protest by a 'black and white' image produced by other students.[30] If we follow Davis' ideas about the cultural history of vision further, one of the implications of a well-understood visual symbolism is that 'styles of depiction—culturally located and historically particular ways of making pictorial representations—have materially affected human visual perception. They constitute what might literally be called *ways of seeing*.'[31] The language and visual language of black and white may literally play a part in enculturating people to see Indigenous and non-Indigenous Australians and their relationships in certain (perhaps rigid and limiting) ways.

But Walker and Wright came together around the metaphor of black and white. Looking at their poetic and personal exchanges through the shadow sister card, it becomes apparent that 'black and white' was a common language that the two women worked with, that made them accessible to each other and that facilitated their communication—something that divided them, but brought them together in recognizing this division.

Wright had first encountered Walker via the manuscript of her first poetry collection, *We are Going*, submitted to her as reader for Jacaranda Press in 1963. She was moved by the passion, anger and persuasive power of the poems and recommended their publication.[32] Wright approached Walker at a writers' function to tell her that 'Son of Mine' was a poem she wished she'd written herself.[33] In the poem, Walker addressed her son Denis with a will to communicate, not the heartbreak or hatred that she knew had characterized so many relations between Aboriginal and non-Indigenous people, but instead the glimpses of the 'brave and fine' that might be found when 'lives of black and white entwine.'[34] For Wright, Walker's vision of deep and equal cross-cultural friendship was like an invitation, one which she dearly wanted to accept.

As they crossed from an admiring acquaintanceship to a fully fledged friendship in about 1969–1972, both women were 'chang[ing] tack to get a fresh breeze.'[35] Walker was preoccupied with race. She had returned from the 1969 World Council of Churches convention on racism in London with a new understanding of what she was fighting for: she had seen herself engaged in a class struggle; she now saw one of race.[36] In a series of passionate speeches and articles that grappled with some of the same dilemmas faced by the American Black Power movement, she argued that Indigenous

Australians must become an empowered, 'unified and solid fighting force' before there could be 'any consideration given to black and white Australians forming a coalition for a better way of life.'[37] Walker was at the head of the move to Aboriginal self-determination and led the formation of a National Tribal Council in which only Indigenous members were to have voting rights,[38] partly inspired by the Black Powerists she had met with in London. Finding that 'black' united *her*, a Noonuccal woman from Stradbroke Island, ground-breaking Aboriginal poet and civil rights campaigner from Queensland, with a potential coalition of races globally, she was elated to think that a global revolution might be at hand, in which the far greater numbers of black people in the world would rise against oppression by the white races.[39]

But Walker 'still had faith in the whites . . . still believed that the whites should be won over.'[40] She involved Wright as she thought through possible approaches to change, sitting at Wright's kitchen table 'on and off for a fortnight' in late 1969 talking out her 'despair over her people's situation . . . rapidly convincing herself that bloody revolution is the only answer.'[41] In the midst of this great personal and political struggle, she trusted Wright as a friend and fellow public thinker, proposing that the two write a book together with Wright 'thinking white' and Walker 'thinking black' to 'guide [Australians] out of the mess they are in.'[42]

From her first volume of poetry, *The Moving Image* (1946), Wright had filled her life's landscapes with disquiet, with histories imperfectly erased like half-buried bones. Not only the earth, but also her own body contained traces of the violence perpetrated against New England's Aboriginal people. Her father's story of 'Nigger's Leap,' a dramatic granite promontory close to her childhood home, where he believed a massacre had been perpetrated, left her with the chilling vision both of denial and of a history that would not disappear from the landscape in the face of it: 'did we not know . . . the black dust our crops ate was their dust?'[43] In 1955, Wright's husband, Jack McKinney, returned from a visit to Palm Island 'distressed by the state of the Aborigines on that miserable island.' This left Wright even more uneasy and wondering what she might do to help.[44] As she came to know Walker, Wright was rethinking what it was she was looking at. The white ascendancy, the white acceptability and white normalcy that she had long been troubled about had become filled with an alarming and at times paralysing guilt, which remained with her until the end of her life. She concluded her half-biography with an apology to 'all the peoples of the old and true Australia on whose land I have trespassed and whom, by being part of my own people, I have wronged . . . I now bend my head and say sorry. Sorry, above all, that I can make nothing right.'[45] But as the 'black' became invested with power and life, it became more of a working contrast: Walker had helped her to find a new place from which to understand the history and the difference which lay between them—'you brought me to you some of the way / and came the rest to meet me.'[46]

Walker's and Wright's black and white was a quest for mutual under-standing. It was completely different from the cursory (but decisive) catego-rization that could type a person according to race with just a flick of the eye over a body—clothing, gait, profile, location—that might denote 'black' or 'white' and then the whole of the person, and thus the whole of that person's relationship to self—the effect of the white gaze that Franz Fanon felt 'fixed' him in a 'crushing objecthood' of blackness.[47]

Crucial to this was another sense. Walker trusted Wright and made audi-ble her history and experience as an Aboriginal woman, and Wright listened (partly metaphorically too, via their correspondence). Jennifer Jones describes their performance of sisterhood as yielding 'a much needed positive model of cross-cultural encounter, a legacy of active listening and learning, a way of liv-ing together in the contact zone.'[48] Jones reworks Henry Reynolds' question 'why weren't we told?' to ask 'why weren't we listening?'—understanding a pervasive 'white deafness' as the obstacle to constructive change in Aborigi-nal–non-Indigenous relations through the latter part of the twentieth century. She holds up Wright's friendship with Walker as a space where real listening happened, for Wright listened deeply and openly—not limiting her listening to a selective hearing which suited a version of Aboriginality that would not require her to change, as so many others had.[49]

The model of listening that Wright offered was open-ended and able to embrace meanings that are becoming increasingly audible to many non-Indigenous Australians in the present. Peter Read has found 'Two Dreamtimes,' and the friendship that it represented, a powerful model for his own friendship with colleague Dennis Foley. Read had researched and written about Aborigi-nal history in NSW for several decades before realizing that the very landscape of his childhood, the northern beaches of Sydney, was within his own lifetime an Aboriginal landscape too. Foley, of Wiradjuri and Gaimarigal descent, grew up among the same rocky ledges and swimming spots, but for him they were mapped by different stories: those of living Aboriginal tradition. Read makes sense of this experience through Wright's poem, calling Foley his 'shadow brother' and adopting Wright's commitment to 'listening . . . with dis-cernment,' a sharing based on 'equal partnership' and through these, a belong-ing to country for both Aboriginal and non-Aboriginal Australians.[50]

The sounds of the two women's voices intermingled within the relation-ship. Walker's voice flowed with a strong current and was often declamatory. Writing to Barbara Blackman, one of her most frequent and intimate cor-respondents, Wright described a visit from Walker in 1980. 'Everyone else seems unspontaneous beside her,' she wrote, '[she] spent most of the time drawing crazy Neo-Aboriginal stuff in sketchbook after sketchbook, writing poems and talking flat out. I couldn't take it as a permanent part of life—one doesn't get a word in.'[51] Walker spoke and wrote with didactic purpose. She had decided that violent protest was not constructive, but instead used her tongue to 'lash' white people, pleased that she had 'hit some sore spots' in her Griffith University address, for example, and driven a number of audience members to walk out, despite the fact that part of her aim in this speech was reconciliatory.[52] Her public friendship with Wright probably played a role in reassuring many non-Indigenous Australians that Walker was acceptable to the audience she most wished to embrace from the 1970s—children, and necessarily their parents and teachers.[53]

As Wright characterized them, Walker's poems asserted an Aboriginal presence at a time when Aborigines were seen as a people of the past, Aboriginal demands when Aboriginal people were understood to be silent and the right of poetry to engage with real life at a time when the Australian scene was dominated by a cultural elite with a narrow conception of the arts as a detached, cerebral pursuit.[54] Walker's often very direct poetic voice jarred with the critics. A 1967 review expressed distaste for her 'shouting' verse and the 'crude thumping rhythms [which] add to the stridency'—'Take care! white racists! / Blacks can be racists too.'[55] She proudly described her own poetry as 'sloganistic, civil-writerish, plain and simple,' although she was also reflective and sensitive in her verse at times.[56] Wright was no less a 'political poet' than her friend, but her poems fell more softly on the ear of the same reviewer (though he also felt they had great effect as fighting poems), partly because of her lyricism, and partly because he could hear her humanity, her appeal to the 'universal' soul and 'love—ordinary human love,' which perhaps resonated with his own soul much more readily than Walker's articulations of a specific, colonized humanity.[57]

The delicate imagery and evocative play of light in Wright's poems, their spontaneity and suppleness, as she characterized human relationships and the landscapes that nourished her, were renowned—Dorothy Porter described her poems as 'shining with meaning.'[58] But the luminous quality of her poems was not reflected in her voice, especially to Wright's own ear. Part of Wright's feeling that she could say or do 'nothing right' as one of the conquerors was a manner that has been described as 'patrician' and her 'Oxford'-like way of speaking.[59] She began an interview with *Meanjin*'s Jim Davidson in 1982 joking about the rather proper voice that would represent her on the recording. She later told him that she felt like she was in the 'wrong camp,' but rejected the notion that she could become Aboriginal via the kind of 'cultural convergence' her fellow poet Les Murray seemed to espouse. Her only hope was reincarnation—'next time' she quipped, 'I propose to be born something rather less British.'[60]

The friends' letters were often casual and quotidian, quite different from the cerebral, formal and symbolic way they communicated about their friendship in public. 'You and I both ought to have permanent knobs on our skulls, the brick walls we keep bashing against,' Wright wrote to Walker in 1970.[61] Wright's occasional offers of financial help to her friend were made in a similar frank, comradely tone, 'I seem to have a bit of a stranglehold on the money situation—so can I help with the bills?'[62] In 1976 Walker wrote to Wright to forewarn her that 'Frank Heimans is doing a film of me. He wants to involve you too. Say no [underlined twice] if you don't want to be involved. I know how busy you are.'

From America, she expressed some of her less glorious feelings about a Fulbright Scholarship visit to Wright: 'all is well with me but I'm getting fat. Too many dinner and lunch parties and horrible food.'[63]

On the shadow sister card, the two women gaze into each other's eyes, in an intersubjective way, in conversation. The conversation was not always harmonious—it was an ongoing conversation, at times awkward and incomplete, and filled with both women's idiosyncrasies. But it was sincere and open-ended. As Brigid Rooney observes, they opened a 'space of risk and revelation' between them, 'a space within which they . . . perhaps began to unfix the immobile, binary identities of coloniser and colonised.'[64]

—

Alison Ravenscroft writes:

> Among critics of whiteness there tends to be a lapse . . . 'white' tends to be taken as a given—as if there are such things as white bodies that are prediscursive and which can be equated with white subjectivity. . . . It is as if there are white subjects (and black subjects) already there, waiting to be found . . . we repeat what we critique.[65]

Ravenscroft extends James Baldwin's thinking to ask whether there are instead only people who desire to be white, people who make themselves white through their reading practices in particular.[66] If the black and white of shadow sisters was a visual metaphor for Walker and Wright to make sense of their own relationship and present it to others, should it also be understood as being at the nexus of their respective desires to be black and white? Was the friendship a locus where Walker made herself black and Wright made herself white?

The shadow sister card depicts the two subjects in symmetry, which enables the contrast between black and white to emerge as a defining difference: the black and white profiles of the shadow sister card literally constitute one another using tonal contrast. Likewise in life, the symmetries Wright and Walker saw in their friendship perhaps provided an opportunity, and a prerequisite, for the foregrounding of black and white. Both poets and public thinkers, the two women found common ground in their strength of commitment to their shared and several causes. Wright wrote to her biographer, 'I wonder why you have never mentioned conservation issues in all the sessions we have had—it's as though you were interested only in the poems that talk about women's life and loves; but those are secondary in my life.'[67] Walker's poem 'My Love' expresses a similar sense of commitment to a quest beyond the merely personal, 'I cannot give / The love that others know, / For I am wedded to a cause.'[68] It was a symmetry of difference perceived and promoted both by the friends themselves and others. Frank Heimans and scriptwriter Jenny Nussinov, applying to the Australian Council for the Arts for funding for their film biography of Walker, *Shadow Sister*, described the film's title as 'self-explanatory in defining the depths of feeling this foremost contemporary Australian poetess has for her black counterpart.'[69] The parity felt within the friendship allowed Walker to exercise her generosity in referring to a converse counterpart-hood on ABC radio in 1990. She told Caroline Jones that Wright was in a flap about being referred to as 'the white Kath Walker' by the 'university set,' whereas she was delighted for this to be said.[70]

Being 'black and white' involved the two women in representing an imagined pan-Aboriginal and pan-settler class. In 'Two Dreamtimes' Wright addressed her friend as a representative of settler-Australians to a representative of the continent's Aboriginal peoples, 'over the desert of red sand / came from your lost country / to where I stand with all my fathers, / their guilt and righteousness.'[71] Walker's 'Sister Poet' responded in the same spirit in its address to Wright and her ' "civilised" kin' from her own place among the 'Koorie dead.'[72]

Though Wright sometimes brushed aside the notion that she might be a prophetic voice for Australians, by the time she was waging her campaign against oil drilling on the Great Barrier Reef in the late 1960s she recognized the 'special advantage' she had in speaking out on conservation issues, and increasingly, Aboriginal rights and reconciliation, as a poet considered ' "the greatest [then] living poet of the Australian landscape", "Australia's greatest woman poet", and even "the greatest woman poet since Sappho." '[73] Walker seemed to feel much less reservation. She declared of her film biography *Shadow Sister*: 'what Kath Walker is is a symbol of the Aboriginal people. I did not make it for Kath Walker but to bring the voices and cries of the Aboriginal people to the surface.'[74] For Walker especially, presenting herself as race-archetypical was not uncontroversial—soon after Walker's death, Eve Fesl, a Gubbi Gubbi Elder from the mainland to the north of Stradbroke Island, gently but firmly corrected Walker's claim to have been the 'voice' of the wider Aboriginal community, reflecting that she had been an 'outsider' in some ways to the Aboriginal perspective, not having borne the brunt of discriminatory laws or experienced life on a mission.[75] Within the context of the friendship, however, this stance allowed Walker and Wright to embrace others in their imagery. A contributor to a collection of poetry in honour of Walker (published in 2000) felt himself involved in reconciliation as he listened to Wright read Walker's poetry; for him 'the thread was joined' and 'a fragile understanding' created.[76]

Coming together in a space where 'black and white' could be focused on by mutual agreement, rather than all the differences of education, boldness, politics, family, generation, gender, et cetera that characterized the women's many other relationships, perhaps allowed them to *be* 'black and white' in a way that they could not elsewhere.

—

The intensive phase of defining the friendship in the early 1970s gave way to a comradely association maintained via frequent correspondence and the occasional visit, but mutual self-definition was something that each turned back to in the early 1990s. The friends were reaching a more reflective phase of life at the same time as public conversations about reconciliation got underway. Walker recalled the story about recognizing herself in Wright's poem 'Canefields Country' in her last major public address, at the Opera House in 1993, and had read 'Two Dreamtimes' on Radio National in 1991.[77] The friendship seemed emblematic to many of those engaged in the promotion of reconciliation, and the symbols of the friendship developed in the 1970s, emerging from their ideas about psychology and race relations, resonated with the reconciliation discourse.[78]

Walker did not think highly of 'reconciliation' as envisaged by the Australian Government, and watching Walker on SBS television in September 1991 confirmed Wright in her view that it was 'a complete con,' which demanded far greater sacrifices from Aboriginal people than from other Australians. As Walker had come to understand the 1967 Referendum over the following decade, reconciliation was likely to make white people feel good about themselves without resulting in real change.[79] Wright quickly made her own views known in public too, stating at the launch of her book *Born of the Conquerors*, 'if I was an Aboriginal person I would be saying JUSTICE BEFORE RECONCILIATION.' She wrote to Walker of the proposed Council for Aboriginal Reconciliation, 'I wouldn't be on it for millions.'[80] But with Walker's death in 1993, those closest to her sent out a message of reconciliation to the world without qualification. Her elder son Denis Walker, speaking at her funeral, was quoted as saying that her legacy was about 'reconciliation and healing,'[81] and Oodgeroo's eldest grandson, Raymond, that 'she wanted us all to live in paradise together . . . and live with each other so the colour of our skin was no different [in significance] to the colour of one's eyes.'[82] Although Wright could see the limits of the reconciliation discourse, she stood beneath its banner, her last public appearance being at a reconciliation march in Canberra in May 2000.[83]

At the same time as providing a model for talking and listening about dispossession, difference and the future, Walker and Wright were also creating a positive metaphorical imagery for their friendship, immersed in the ongoing problems of seeing in black and white at the same time as being a collegial, respectful and warm vision of black and white. The 'shadow sister' card, as one artist's interpretation of the friends' verbal visuality of race, may be seen as a predecessor of the photographic products of the 'Dare to Lead' project, and images like the 1996 Council of Australian Reconciliation poster, 'Together we can't Lose,' showing athletes (and arch-rivals) Melinda Gainsford and Cathy Freeman embracing as they wear the Australian and Aboriginal flags. With images like Mervyn Bishop's highly influential 1975 photograph of Prime Minister Gough Whitlam pouring a handful of earth into traditional owner Vincent Lingiari's hands, Walker and Wright's mostly verbal–visual imaging of friendship may have helped inspire the visuality of reconciliation.

Notes

1 Papers of Judith Wright, National Library of Australia Manuscript Collection, MS 5781, Box 74, Folder 533, card signed 'JS '93'.
2 McKinney, 'Wyndham, George (1801–1870)', *ADB*.
3 Clarke and McKinney 2006, pp. 2–4.
4 Judith Wright, interview with Jim Davidson, 26 May 1982, Meanjin Collection, National Library of Australia, Tape 1, side A.
5 Hanf 2011.
6 Cochrane 1994, p. 19.
7 Bonyhady 2007, p. 20.
8 Papers relating to Oodgeroo Noonuccal, 1991–2000 (Kathie Cochrane), Fryer Library Manuscript Collection, UQFL 286, Wright to Cochrane, 2 October 1991.
9 Kath Walker, interview with Hazel de Berg, Hazel de Berg Collection, National Library of Australia, DeB923, 10 March 1976, side A; Darling 1998, p. 187.
10 Judith Wright, interview with Ramona Koval for ABC Radio National, 'Books and writing' Program, 30 June 2000.
11 Cochrane 1994, pp. 43–48.
12 Clarke and McKinney 2006, p. xii.
13 This friendship and its public performance are explored in much greater depth in my doctoral thesis: Dortins 2012.
14 Papers of Judith Wright, Box 74, Folder 533, Judith Sercombe to Wright undated, and cover letter 16 September 1993.
15 Papers of Judith Wright, Box 74, Folder 533, see for example (among those who were close to Wright), Dymphna Clark to Wright, 16 September 1993; Isobel White to Wright, 17 September 1993.
16 Papers of Judith Wright, Box 74, Folder 533, Felicity to Wright, undated.
17 Kath Walker, interview with Hazel de Berg, side B; Wright 1955, p. 26; Walker 1992, pp. 78–80.
18 Wright 2003, pp. 315–18. Permission to quote from the poem courtesy Harper-Collins Publishers.
19 See for example, Ryan 1999, p. 29; King-Smith 2007, pp. 125–27.
20 Papers relating to Oodgeroo Noonuccal, 1991–2000 (Kathie Cochrane), Wright to Cochrane, undated; Bull 1997.
21 Heimans 1977.
22 Jung 1959, pp. 8–10. See Wright to Barbara Blackman, 22 April 1958 and Wright to Gillian Coote, 1989 in Clarke and McKinney 2006, pp. 109, 216, 450.
23 Jung 1959, pp, 8–10; Judith Wright, interview with Jim Davidson, 26 May 1982, Tape 1, side A.
24 Jones 2003, p. 47.
25 Grandlin 2015.
26 Walker 1980, pp. 169–70.
27 Kath Walker, 'Sister poet', cited in Read Lauer 1978, pp. 89–90. For an exploration of the relationship between the two poems, see Dortins 2013, pp. 166–77.
28 Davis 2011, p. 8.
29 Paradies 2006, p. 359.
30 Huggins c2006, pp. 6–7.
31 Davis 2011, p. 6.
32 Wright, 'The poetry' in Cochrane 1994, pp. 165–69.
33 Kath Walker, interview with Julianne Schwenke, Fryer Library collection, Tape 4, side 2, undated, c1978.
34 Walker 1964, p. 13.
35 Collins 1994, p. 18.
36 Taffe 2007–08.

37 Walker 1969a, p. 2; Walker 1969b, p. 6; Ture and Hamilton 1992, pp., 202–06.
38 Read 1990, p. 79.
39 Papers of Oodgeroo Noonuccal, Fryer Library Manuscript Collection, UQFL 84, Box 30, 'Report to the Australian Council of Churches on the World Council of Churches "Consultation on Racism" ' held in London, May 1969.
40 Kath Walker, interview with Julianne Schwenke, Tape 4, side A.
41 Wright to Kathleen McArthur, undated; Clarke and McKinney 2006, p. 203.
42 Papers of Judith Wright, Box 64, Folder 470, Walker to Wright, 8 September 1969.
43 Wright 1946, p. 23.
44 Wright 1999, p. 284.
45 Ibid., p. 296.
46 Wright, 'Two dreamtimes', 2003, pp. 315–18.
47 Fanon 2001, p. 86, cited by Morris.
48 Jones 2003, p. 49.
49 Ibid., p. 44.
50 Read 2000, pp. 21–29, 198–224.
51 Clarke and McKinney 2006; Wright to Blackman, 18 November 1980, 339.
52 Papers of Judith Wright, Box 64, Folder 471, Walker to Wright, 13 September 1989.
53 Walker wrote in a postcard to Wright from Moongalba, 'I'll concentrate on lecturing children. After all they haven't yet been brainwashed to the extent of their adults who are beyond help in view of being so mentally constipated', Papers of Judith Wright, Box 64, Folder 470, Walker to Wright, 10 February 1977. See also Kath Walker, interview with Hazel de Berg, side B.
54 Wright, 'The poetry', in Cochrane 1994, pp. 166–69.
55 Lee 1967, p. 64; Walker 1970, p. 85.
56 Walker 1975, p. 39.
57 Lee 1967, p. 66.
58 Dorothy Porter quoted in ABC Radio National 'Poetica' program: 'Shining with meaning: The poetry of Judith Wright', produced by Anna Messeriti, broadcast 23 June 2012.
59 Capp 2009.
60 Judith Wright, interview with Jim Davidson, Tape 1, side A.
61 Papers of Oodgeroo Noonuccal, Box 39, Wright to Walker, 4 August 1970.
62 Papers of Oodgeroo Noonuccal, Box 54, Wright to Walker, 18 June 1980.
63 Papers of Judith Wright, Box 64, Folder 470, Walker to Wright, 22 July 1976 and 21 November 1978.
64 Rooney 2009, p. 72.
65 Ravenscroft 2007.
66 Ibid.
67 Clarke and McKinney 2006; Wright to Veronica Brady, 17 January 1996, 528.
68 Walker 1970, p. 50.
69 Nussinov, 'Exploratory points for the film 'Shadow sister', May 1976, from the collection of Frank Heimans.
70 Oodgeroo Noonuccal, interview with Caroline Jones c1990, 'A search for meaning' series, ABC Radio Tapes, Tape 181, side B.
71 Wright 2003, pp. 315–18.
72 Walker, 'Sister poet', cited in Read Lauer 1978, pp. 89–90.
73 Bonyhady 2007, p. 20, quoting partly from Wright's *The coral battleground*, 1977.
74 *Sydney Morning Herald*, 1 August 1977, p. 4. See also Davidson 1977, pp. 428–29. It was a stance that reflected her experience as one of the few accomplished Aboriginal speakers available to take the lectern in the Referendum campaign, and also her reception as one of the first published Aboriginal poets, but perhaps seemed to claim unwarranted authority as she continued to make these claims of her own work into the 1970s and 1980s.

75 Fesl 1994, pp. 143–44. See also Cochrane 1994, p. 72.
76 Robertson 2000, p. 83.
77 Walker, 'Writers of Australia, "I dips me lid" ' in Cochrane 1994, pp. 218–19; Oodgeroo Noonuccal, interview with Caroline Jones, side B.
78 See for example Gallagher 2005, pp. 37–48.
79 Papers of Judith Wright, Box 74, Folder 534, Wright to Walker, 4 September 1991; Kath Walker, interview with Hazel de Berg, side B.
80 Papers of Judith Wright, Box 74, Folder 534, Wright to Walker, 4 September 1991.
81 Fiona Hamilton 'Tributes flow for "grand old lady" ', newsclipping in Papers of Judith Wright, Box 74, Folder 533.
82 'Farewell: May her Dreaming be powerful', *Courier Mail*, 21 September 1993.
83 *The Canberra Times*, 29 May 2000, p. 6; Huggins 2001.

References

Bonyhady, T., 2007. 'Torn between art and activism', *Local-Global: Identity, Security, Community*, 3, p. 20.
Bull, S., 1997. 'Vigil for justice', *Green Left Weekly*, 277, 4 June.
Capp, F., 2009. 'In the garden: Judith Wright and Nugget Coombs', *The Monthly*, 46, June.
Clarke, P. and McKinney, M., 2006. *With love and fury: Selected letters of Judith Wright*, Canberra: National Library of Australia.
Cochrane, K., 1994. *Oodgeroo*, Brisbane: University of Queensland Press.
Collins, J., 1994. 'A mate in publishing', in A. Shoemaker, ed., *Oodgeroo: A tribute*, Brisbane: Australian Literary Studies and University of Queensland Press, p. 18.
Darling, E., 1998. *They spoke out pretty good: Politics and gender in the Brisbane Aboriginal rights movement 1958–1962*, Melbourne: Janoan Media Exchange.
Davidson, J., 1977. 'Interview—Kath Walker', *Meanjin*, 36.4, pp. 428–29.
Davis, W., 2011. *A general theory of visual culture*, Princeton: Princeton University Press.
Dortins, E., 2012. 'The lives of stories: Making histories of Aboriginal-settler friendship', Thesis, University of Sydney.
Dortins, E., 2013. 'Apology and absolution—The poetic and public dialogue of Judith Wright and Kath Walker', in T. Clark and S. Henriss-Andersse, eds., *Testimony, witness, authority: The politics and poetics of experience*, Newcastle upon Tyne, UK: Cambridge Scholars Publishing, pp. 166–77.
Fesl, E., 1994. 'The road ahead', in A. Shoemaker, ed., *Oodgeroo: A tribute*, Brisbane: Australian Literary Studies and University of Queensland Press, pp. 143–44.
Gallagher, K., 2005. 'Towards reconciliation: Inspiration and leadership—Oodgeroo of the Tribe Noonuccal (1920–1993) and Judith Wright (1915–2000)', *Agenda*, 41.1–2, pp. 37–48.
Grandlin, G., 2015. 'Blackness as being, whiteness as nothingness', *The Nation*, 23 June.
Hanf, R., 2011. 'Native title recognition for the Quandamooka People', National Native Title Tribunal website, news page, 4 July.
Heimans, F., 1977. *Shadow sister: A film biography of Australian Aboriginal poet Kath Walker*, Sydney: Cinetel Productions.
Huggins, J., 2001. Judith Wright address, Native Title Representatives Bodies Legal Conference, Townsville, 28–30 August.
Huggins, J., 2006. *Images of reconciliation*, Dare to Lead Conference, 18 November 2005, Reconciliation Australia and Department of Education Science and Training.Jones, J., 2003. 'Why weren't we listening? Oodgeroo and Judith Wright', *Overland*, 171, Winter, p. 47.

Jung, C.G., 1959. *Aion: Researches into the phenomenology of the self*, translated by R.F.C. Hall in Collected Works of C.G. Jung series, vol. 9, part 2, London: Routledge and Kegan, pp. 8–10.

King-Smith, S., 2007. 'Ancestral echoes: Spectres of the past in Judith Wright's poetry', *Journal of the Association for the Study of Australian Literature*, Special Issue, pp. 125–27.

Lee, S.E., 1967. 'Poetic fisticuffs', *Southerly*, 27, p. 64.

Morris, R., 2001. 'Reading photographically: Translating whiteness through the eye of the empire', *Hecate*, 27.2, pp. 86–96.

Paradies, Y.C., 2006. 'Beyond black and white: Essentialism, hybridity and indigeneity', *Journal of Sociology*, 42.4, pp. 355–67.

Ravenscroft, A., 2007. 'Who is the white subject? Reading, writing, whiteness', *Australian Humanities Review*, 42, August–September.

Read, P., 1990. 'Cheeky, insolent and anti-white: The split in the Federal Council for the advancement of Aboriginal and Torres Strait Islanders—Easter 1970', *Australian Journal of Politics & History*, 1, p. 79.

Read, P., 2000. *Belonging*, Cambridge: University of Cambridge Press.

Read Lauer, M., 1978. 'Kath Walker at Moongalba: Making a new dreamtime', *World Literature Written in English*, 17.1, pp. 89–90.

Robertson, J., 2000. 'The two sisters', in J. Evans ed., *Moongalba: Poems in honour of Oodgeroo*, Brisbane: IBIS Editions Australien, p. 83.

Rooney, B., 2009. *Literary activists: Writer-intellectuals and Australian public life*, Brisbane: University of Queensland Press.

Ryan, G., 1999. 'Uncertain possession: The politics and poetry of Judith Wright', *Overland*, 154, p. 29.

Taffe, S., n.d. 'Kath Walker, Collaborating for indigenous rights 1957–1973' website, National Museum of Australia, created 2007–08. Available at http://indigenousrights.net.au/person.asp?pID=988, accessed 28 August 2009.

Ture, K. and Hamilton, C.V., 1992. 'Afterword', in *Black power: The politics of liberation in America*, New York: Vintage Books, pp. 202–06.

Walker, K., 1964. 'Son of mine', in *We are going*, Brisbane: The Jacaranda Press, p. 13.

Walker, K., 1969a. 'Racism: Double-thinking, complex state of mind', *Origin*, 1.1, August, p. 2.

Walker, K., 1969b. 'Black-white coalition can work', *Origin*, 18, September, p. 6.

Walker, K., 1970. 'Racism' and 'My Love', in *My people: A Kath Walker collection*, Brisbane: Jacaranda Press, p. 85.

Walker, K., 1975. 'Aboriginal literature', *Identity*, 2.3, p. 39.

Walker, K., 1992. *Stradbroke dreamtime*, Sydney: Angus and Robertson.

Walker, S., 1980. *The poetry of Judith Wright: A search for unity*, Melbourne: Edward Arnold.

Wright, J., 1946. 'Nigger's Leap, New England' and 'South of my days', in *The moving image*, Melbourne: The Meanjin Press, pp. 23 and 28.

Wright, J., 1955. 'Cane country', in *The two fires*, Sydney: Angus and Robertson, p. 26.

Wright, J., 1999. *Half a lifetime*, Melbourne: Text Publishing.

Wright, J., 2003. 'Two dreamtimes', in *Collected Poems*, Sydney: HarperCollins, pp. 315–18.

Wright McKinney, J., n.d. 'Wyndham, George (1801–1870)', *Australian Dictionary of Biography*, Canberra: Australian National University. Available at http://adb.anu.edu.au/biography/wyndham-george-2824/text4049, published first in hardcopy 1967, accessed 16 January 2016.

Contributors

Editors

Joy Damousi is Professor of History at the University of Melbourne and an Australian Research Council Kathleen Fitzpatrick Laureate Fellow. Her publications include *The Labour of Loss: Mourning, Memory and Wartime Bereavement in Australia* (Cambridge University Press, 1999), a study of grief, war and memory; *Colonial Voices: A Cultural History of English in Australia, 1840–1940* (Cambridge University Press, 2010), a history of sound, voice and speech; and *Talking and Listening in the Age of Modernity: Essays on the History of Sound* (ANU Press, 2007) (ed. with Desley Deacon). She is currently researching *'Hell Sounds': The Soundscape of War, 1914–1945*, an examination of sound and the two world wars of the twentieth century.

Paula Hamilton is a cultural historian who explores the links between personal and public memories and has published mainly in oral history and memory studies. She is currently an adjunct Professor of History at the University of Technology. Her first published essay on the senses was 'The Proust Effect: Oral History and the Senses' in *Oxford Oral History Reader* edited by Don Ritchie (Oxford University Press, 2010), which has been republished in the third edition of Robert Perks and Alistair Thomson (eds), *The Oral History Reader* (Routledge, 2016). She has also recently edited a special issue of *Women's History Review* (with Mary Spongberg) on 'Women's History and the Digital Turn' (forthcoming 2016) and co-edited (with Jim Gardner) *The Oxford Handbook of Public History* (forthcoming 2016).

Chapter Authors

Jennifer Bowen is currently completing a PhD at the University of Melbourne on the cultural implications of radio talks in Australia. This follows an international career as a radio producer: she was a senior producer with the BBC World Service Features and Arts Department, during which time she was shortlisted for a Sony Award, and while working for the Australian Broadcasting Commission she was an award winner at the New York International Radio Festival in 2011. She has taught radio journalism at Monash University and undertook the ABC Radio National production for the 2014 Australian Generations oral history project.

Diane Collins was director of the Historical and Cultural Studies programme at the Sydney Conservatorium of Music, University of Sydney, before retirement in July 2010. She has written on the history of high and low cultural institutions in Australia. *Sounds from the Stables: The Story of the Sydney Conservatorium* was published in 2001. More recently, she has researched and published on diverse aspects of the auditory history of Australia. Her article, 'The "Voice" of Nature? Kookaburras, Culture and Australian Sound' won the 2011 Barrett Prize for the best article published in the *Journal of Australian Studies*.

Kate Darian-Smith holds concurrent positions as Professor of Australian Studies and History and Professor of Cultural Heritage at the University of Melbourne. She has published widely on memory and commemoration, including the landmark *Memory and History in Twentieth Century Australia* (Oxford University Press, 1994; ed. with Paula Hamilton); on the histories of childhood and play, including *Children, Childhood and Cultural Heritage* (Routledge, 2013; ed. with C. Pascoe); and on imperialism and nationhood, most recently in *Conciliation on Colonial Frontiers: Conflict, Performance and Commemoration in Australia and the Pacific Rim* (Routledge, 2015, ed. with P. Edmonds).

Emma Dortins works for the Heritage Division of the NSW Office of Environment and Heritage. She leads a small team listing significant places on the State Heritage Register and helping Aboriginal people to nominate places with cultural importance to them to be recognized as Aboriginal Places under the *National Parks and Wildlife Act*. She has a PhD in history from the University of Sydney for her thesis 'The Lives of Stories: Making Histories of Aboriginal-Settler Friendship,' where she explored popular, public, family and local history-making by following the lives of four stories told and retold across many decades.

David Goodman teaches US history in the School of Historical and Philosophical Studies at the University of Melbourne. He holds a PhD in History from the University of Chicago and is the author of *Radio's Civic Ambition: American Broadcasting and Democracy in the 1930s* (Oxford University Press, 2011). He is currently working on a history of fortune-telling in modern America and a grass-roots history of the debate about US entry into World War II.

Vannessa Hearman is a lecturer in Indonesian Studies at Charles Darwin University. Her research interests include the 1965–66 anti-communist violence in

Index